Nordic Wittgenstein Studies

Volume 4

The series publishes high-quality studies of Ludwig Wittgenstein's work and philosophy. It is affiliated with The Nordic Wittgenstein Society, The Wittgenstein Archives at the University of Bergen and The von Wright and Wittgenstein Archives at the University of Helsinki. The series welcomes any first rank study of Wittgenstein's philosophy, biography or work, and contributions in the subject areas of philosophy and other human and social studies (including philology, linguistics, cognitive science and others) that draw upon Wittgenstein's work. It also invites studies that demonstrate the philosophical relevance of Wittgenstein's Nachlass as well as purely philological or literary studies of the Nachlass. Each submission to the series, if found eligible by the series editor, is peer reviewed by the editorial board and independent experts. The series accepts submissions in English of approximately 80 000 – 125 000 words. For further information (about how to submit a proposal, formatting etc.), please contact: niklas.forsberg@filosofi.uu.se.

More information about this series at http://www.springer.com/series/13863

Benjamin De Mesel

The Later Wittgenstein and Moral Philosophy

Springer

Benjamin De Mesel
KU Leuven
Leuven, Belgium

ISSN 2520-1514 ISSN 2520-1522 (electronic)
Nordic Wittgenstein Studies
ISBN 978-3-030-07379-4 ISBN 978-3-319-97619-8 (eBook)
https://doi.org/10.1007/978-3-319-97619-8

Cover Illustration: The cover makes use of Wittgenstein Nachlass MS 115, page 118. The Master and Fellows of Trinity College Cambridge and the University of Bergen have kindly permitted the use of this picture.

This Springer imprint is published by the registered company Springer Nature Switzerland AG
The registered company address is: Gewerbestrasse 11, 6330 Cham, Switzerland

Acknowledgements

First of all, I would like to thank Stefan Rummens. Ever since I started this project, Stefan has been approachable, supportive and confident in its success. His MA course on Wittgenstein and ethics sparked my interest in the topic of this book, and its central idea of applying Wittgenstein's later methods to problems in moral philosophy is his. He has read and encouragingly commented upon (sometimes several versions of) almost all of the chapters. I have discussed parts of this book with many people, and every chapter has received written comments by several anonymous reviewers, but however helpful these comments have been, they never matched the invariably outstanding quality of Stefan's remarks. These remarks were insightful and always constructive, aimed at making the chapter better in every possible way.

Together with Stefan Rummens, James M. Thompson, Sybren Heyndels, Audun Bengtson and Dimitris Gakis, I organized three conferences in Leuven entitled 'Wittgensteinian Approaches to Moral Philosophy'. I am grateful to the invited speakers (Sophie-Grace Chappell, Anne-Marie Christensen, James Conant, Alice Crary, Raimond Gaita, Hans-Johann Glock, Edward Harcourt, Julia Hermann, Oskari Kuusela, Sabina Lovibond, Michel Meliopoulos, Danièle Moyal-Sharrock, Nigel Pleasants, Genia Schönbaumsfeld, Joachim Schulte, Martin Stokhof and Walter Van Herck) for accepting our invitation, and to all the other speakers and participants for bringing outstanding Wittgenstein scholarship to Leuven. I would like to thank Stefan, James, Sybren, Audun and Dimitris for their help with the organization of the conferences.

Over the last 5 years, I have presented papers at several conferences: the Wittgenstein conferences in Leuven; the Institute of Philosophy Graduate Student Conference (Leuven); the Contemporary Significance of Ordinary Language Philosophy (Turku, Finland); the NPPW-CESPP Conference (Vaalbeek); Mind, Language and Action (36th International Wittgenstein Symposium) (Kirchberg am Wechsel, Austria); Dutch Research School of Philosophy (OZSW) 2013 Conference (Rotterdam, The Netherlands); Analytical and Continental Philosophy: Methods and Perspectives (37th International Wittgenstein Symposium) (Kirchberg am Wechsel, Austria); Realism – Relativism – Constructivism (38th International

Wittgenstein Symposium) (Kirchberg am Wechsel, Austria); Dutch Research School of Philosophy (OZSW) 2015 Conference (Amsterdam, The Netherlands). I am indebted to the audiences of these conferences for instructive comments on my presentations.

I have participated in three Ludwig Wittgenstein Summer Schools in Kirchberg am Wechsel (Austria). I must thank the instructors (James Conant, Cora Diamond, P.M.S. Hacker, Joachim Schulte) for deepening my understanding of Wittgenstein's thought. Almost all the chapters of this book, as well as other aspects of Wittgenstein's work, have been amply discussed with participants of the Wittgenstein Summer Schools or the International Wittgenstein Symposia in Kirchberg. I would like to thank, in particular, Mindaugas Gilaitis, Nicole Hausen, Marco Nuzzaco, Désirée Weber, Stefanie Dach, Hannah Read, Daniel Sharp, Wim Vanrie and Sybren Heyndels.

I was a Recognised Student at the University of Oxford in the first semester of the academic year 2014–2015. I would like to thank Edward Harcourt for being my supervisor there and for commenting upon my work, and Sabina Lovibond and Sebastian Grève for several meetings in which we discussed Wittgenstein and moral philosophy. On a non-academic note, I am grateful to Iris Geens, Robert Esnouf and Aidan Esnouf for being perfect hosts.

All these adventures abroad notwithstanding, most of the research for this book was conducted in Belgium. Special thanks go to the Institute of Philosophy (KU Leuven) for providing an excellent research environment. Attending the Ethics Reading Group has substantively enlarged my knowledge and understanding of the field, and I would like to thank its members (Arnold Burms, Bart Pattyn, Xavier Vanmechelen, Jef Peeters, Bart Engelen, Chris Bessemans, Anne Polkamp, Jinzhou Ye, Michael Jewkes) for encouragement and comments on some of my work.

Not only the academics at the Institute are to be thanked, however. Fran Venken and Sofie Keyaerts offered much appreciated help in organizing the Wittgenstein conferences in Leuven. Ingrid Dierickx and Sofie Keyaerts took meticulous care of the reimbursements of costs for books, travel and conferences. I cannot emphasize enough the crucial role played by the library and its staff (Steven Spileers, Elke Caubergh, Relinde Geys and Jorinde Geeraert). Having been a student at the Institute, I got used to finding almost every book that I needed in the library. Contacts with academics abroad have made it clear to me how exceptional that situation is and how privileged the Institute's philosophers are. Another respect in which they are privileged concerns the availability of a proofreader for those who are not native speakers of English. Without the generous, careful and precise editing work of Sean Winkler and Jeremy Hovda, readers of this book would have encountered numerous linguistic blunders.

I must also thank two rather invisible groups of people: first, the anonymous reviewers of my articles, for their time and (sometimes) helpful comments; second, the FWO (Research Fund – Flanders), for their trust and money and for not burdening me with excessive administrative duties.

I am grateful to Niklas Forsberg, editor of Nordic Wittgenstein Studies, for his support, to two anonymous reviewers of the manuscript and to all the people at Springer who were involved in the production process.

Finally, I would like to express my gratitude to my friends, my parents and my wife Lien. It is impossible to convey in a few lines what their love and support have meant to me. I can only say thank you.

Contents

1 Introduction 1
1.1 Wittgenstein's Ethical Views 2
1.1.1 The Problem of Scarcity 2
1.1.2 How to Get Clearer on Wittgenstein's Ethical Views 3
1.2 Wittgenstein's Ethical Views and His Later Philosophy 5
1.3 Wittgenstein's Later Philosophy and Contemporary Moral Philosophy 9
1.4 Structure and Overview of the Book 14
1.4.1 Part I. Wittgenstein's Later Philosophy 14
1.4.2 Part II. Wittgenstein's Later Philosophy and His Ethical Views 17
1.4.3 Part III. Wittgenstein's Later Philosophy and Contemporary Moral Philosophy 20
References 24

Part I Wittgenstein's Later Philosophy

2 On Wittgenstein's Comparison of Philosophical Methods to Therapies 31
2.1 Introduction 31
2.2 The Illness: Are Philosophical Questions Illnesses? 33
2.3 The Patient and the Therapist: Are They Philosophers? 36
2.4 The Therapies: Are They Psychological? 39
2.5 The Ideal of Health: The End of Philosophy? 43
2.6 Conclusion 45
References 46

3 Surveyable Representations, the 'Lecture on Ethics', and Moral Philosophy 49
3.1 Introduction 49
3.2 The Concept of a Surveyable Representation: Two Interpretations 51

3.3 The Concept of a Surveyable Representation: A Third Interpretation 54
3.4 The 'Lecture on Ethics' as a Surveyable Representation 58
3.5 Are Surveyable Representations Useful for Contemporary Moral Philosophy? 63
References 67

Part II Wittgenstein's Later Philosophy and His Ethical Views

4 Wittgenstein, Meta-ethics, and the Subject Matter of Moral Philosophy 71
4.1 Introduction 71
4.2 'Ethics Has No Particular Subject Matter' 72
4.3 'There Is and Can Be No Such Thing as Meta-ethics' 77
4.4 'A Classification by Use Rather than by Subject Matter' 81
4.5 'Ethics Is a Pervasive Dimension of Life Rather than a Distinguishable Region or Strand of It' 85
4.6 Conclusion 88
References 89

5 Wittgenstein and Objectivity in Ethics. A Reply to Brandhorst 91
5.1 Introduction 91
5.2 Brandhorst's Arguments 92
5.2.1 In a Different Way, Different Kinds, in a Different Sense 93
5.2.2 No View from Nowhere 93
5.2.3 There Is a Language-Game in Which the Expression Has a Use 94
5.2.4 A Deflationary Conception, No Metaphysical Depth 94
5.2.5 With a Certain Conception of X, We Also Obtain a Conception of a Related Notion Y 94
5.2.6 The Analogy to Logic and Mathematics 95
5.3 Arguments for Objectivity? 96
5.3.1 In a Different Way, Different Kinds, in a Different Sense 96
5.3.2 No View from Nowhere 96
5.3.3 There Is a Language-Game in Which the Expression Has a Use 98
5.3.4 A Deflationary Conception, No Metaphysical Depth 98
5.3.5 With a Certain Conception of X, We Also Obtain a Conception of a Related Notion Y 99
5.3.6 The Analogy to Logic and Mathematics 100
5.4 Brandhorst's Conception of Objectivity 101
5.5 How Brandhorst Reads Wittgenstein on Objectivity 104
5.6 Conclusion 109
References 109

Part III Wittgenstein's Later Philosophy and Contemporary Moral Philosophy

6 Seeing Colour, Seeing Emotion, Seeing Moral Value . . . 113
6.1 Introduction . . . 113
6.2 Seeing Colour, Seeing Emotion, Seeing Moral Value: Similarities . . . 115
6.3 The Active Element . . . 118
6.4 Education and Concept-Mastery . . . 120
6.5 Blindness . . . 122
6.6 Appropriate Perceivers and Normal Observation Conditions . . . 124
6.7 Conclusion . . . 127
References . . . 129

7 Do Moral Questions Ask for Answers? . . . 131
7.1 Introduction . . . 131
7.2 Why Is S2 Absurd? . . . 132
7.3 Why S2 Is Absurd . . . 136
7.3.1 Jonathan's Question . . . 136
7.3.2 Mary's Answer . . . 138
7.3.3 Jonathan's Reaction . . . 142
7.4 Moral Questions and Narrow Answers . . . 144
References . . . 150

8 How Morality Can Be Absent from Moral Arguments . . . 153
8.1 Introduction . . . 153
8.2 Denying a Moral Certainty . . . 154
8.3 No 'Attitude Towards a Soul' . . . 160
8.4 No Deepened Understanding . . . 163
8.5 Ignoring the Personal . . . 166
8.6 An Absence of Morality . . . 170
References . . . 172

9 Conclusion . . . 175
References . . . 181

Index . . . 183

Original Places of Publication

This book is a collection of articles. Each of the three parts of the book consists of two or three chapters, and each chapter (except for Chaps. 1 and 9, the introduction and the conclusion) has been published as an independent journal article. I have imposed uniformity of reference style upon the papers in this collection. Many of the papers have been amended in order to correct errors, to add qualifications or further clarifications.

The original publication locations are as follows:

Chapter 2

De Mesel Benjamin. 2015. On Wittgenstein's Comparison of Philosophical Methods to Therapies. *International Journal of Philosophical Studies* 23: 566–583. Published by Taylor & Francis. See http://www.tandfonline.com/doi/abs/10.1080/09672559.2015.1034749.

Chapter 3

De Mesel Benjamin. 2014. Surveyable Representations, the 'Lecture on Ethics', and Moral Philosophy. *The Nordic Wittgenstein Review* 3: 41–69. See http://www.nordicwittgensteinreview.com/article/view/1922.

Chapter 4

De Mesel Benjamin. 2015. Wittgenstein, Meta-Ethics and the Subject Matter of Moral Philosophy. *Ethical Perspectives* 22: 69–98. Published by Peeters Publishers. See http://poj.peeters-leuven.be/content.php?id=3073458&url=article.

Chapter 5

De Mesel Benjamin. 2017. Wittgenstein and Objectivity in Ethics. A Reply to Brandhorst. *Philosophical Investigations* 40: 40–63. Published by Wiley-Blackwell. See http://onlinelibrary.wiley.com/doi/10.1111/phin.12129/abstract.

Chapter 6

De Mesel Benjamin. 2016. Seeing Color, Seeing Emotion, Seeing Moral Value. *The Journal of Value Inquiry* 50: 539–555. Published by Springer. See https://link.springer.com/article/10.1007/s10790-015-9535-4.

Chapter 7

De Mesel Benjamin. 2015. Do Moral Questions Ask for Answers? *Philosophia. Philosophical Quarterly of Israel* 43: 43–61. Published by Springer. See https://link.springer.com/article/10.1007/s11406-014-9565-3.

Chapter 8

De Mesel Benjamin. 2016. How Morality Can Be Absent from Moral Arguments. *Argumentation* 30: 443–463. Published by Springer. See https://link.springer.com/article/10.1007/s10503-015-9389-8.

I am grateful to the publishers of these journals for permission to reprint my articles.

Chapter 1
Introduction

This book is a collection of articles, not a monograph. There is not one master argument; rather, the chapters engage, each in their own way, with contemporary debates about the later Wittgenstein's philosophical methods and/or the relevance of Wittgenstein's later thought for moral philosophy. The chapters have been published as independent journal articles and can be read on their own. Taken together, they contribute to the overall aim of this book, which is to show that Wittgenstein's later methods can fruitfully be applied to several problems in moral philosophy. That may seem to be a strange claim, because Wittgenstein's contribution to the discipline of philosophy is most often situated in logic, the philosophy of language, the philosophy of mind and the philosophy of mathematics. Why then think that his thought is relevant for moral philosophy?[1]

In Sect. 1.1, I will argue that Wittgenstein's explicitly stated ethical views are so short and scarce that they can hardly be regarded as constituting, in themselves, a significant contribution to moral philosophy.[2] However, that does not mean that his ethical remarks are philosophically irrelevant. I will contend in Sect. 1.2 that Wittgenstein's later philosophy can serve as a toolbox with the help of which some of Wittgenstein's ethical views can be elucidated. Section 1.3 is devoted to showing that insights from Wittgenstein's later philosophy can also be applied to problems in moral philosophy that have no obvious link with Wittgenstein's ethical views. In Sect. 1.4, I present an overview of the different articles in this book with particular

[1] Books that explore themes related to those treated here include (but are not limited to) Barrett (1991); Cavell (1999); Crary (2007a, b, 2016); Diamond (1991); Edwards (1982); Gaita (2002, 2004); Hermann (2015); Iczkovits (2012); Johnston (1989, 1999); Litwack (2009); Lovibond (1983, 2002); O'Connor (2008); O'Hara (2018); Rhees (1969, 1999); Tilghman (1991); Winch (1972, 1987); Wisnewski (2007). There are two journal special issues that focus on Wittgenstein's views on ethics and on Wittgensteinian ethics (Gleeson 2002; De Mesel and Thompson 2015). Recent collections of articles on Wittgenstein and moral philosophy are Agam-Segal and Dain (2018); Kuusela and De Mesel (forthcoming).

[2] I will use 'ethics' and 'moral philosophy', 'ethical' and 'moral' interchangeably.

B. De Mesel, *The Later Wittgenstein and Moral Philosophy*, Nordic Wittgenstein Studies 4, https://doi.org/10.1007/978-3-319-97619-8_1

attention to the overall structure, the ways in which the chapters are linked and their Wittgensteinian character.

1.1 Wittgenstein's Ethical Views

1.1.1 The Problem of Scarcity

Although Wittgenstein is not usually regarded as a moral philosopher, his work does contain explicit remarks on ethics. One could think, therefore, that the importance of Wittgenstein's thought for moral philosophy is to be found, in the first place, in these remarks.

This proposal is problematic because Wittgenstein's explicit remarks on ethics are short, scarce and scattered. His only sustained treatment of ethics, 'A Lecture on Ethics' (Wittgenstein 2014), is hardly ten pages long. Other remarks can be found in the *Notebooks* (Wittgenstein 1979a: 77–80, 84, 91) and in the *Tractatus* (Wittgenstein 2005a: §§6.4–6.5).[3] The *Philosophical Investigations* are almost silent about the topic (if Wittgenstein 2009a: §77 and §304 warrant the use of 'almost' here). Wittgenstein further touches upon ethics and moral problems in some of his lectures (Wittgenstein 1966: 3, 70; Wittgenstein 1993: 103, 106), some remarks in the *Nachlass* (MS 130, 7; MS 131, 8–9; MS 136, 47b; MS 140, 33; TS 219, 11) and in personal notes (Wittgenstein 1998: 5, 40, 93, 95), diaries (Wittgenstein 2003: 77, 85, 117, 223), letters (Wittgenstein 2003: 321) and conversations with friends and colleagues (see, for example, Bouwsma 1986; Malcolm 1984; Rhees 1965; Waismann 1965).

The fact that Wittgenstein wrote very little about ethics means that any philosophical treatment of Wittgenstein's ethical views is immediately confronted with a number of problems. First, there are questions of continuity. Did Wittgenstein's view on ethics change? If so, why was he not explicit about that in the *Investigations?* Did he assume that it was obvious that he had abandoned his earlier views on ethics and that he had adopted a novel view of ethics consonant with his notions of language-games and forms of life, so that it was not necessary to talk about it explicitly?[4] In my view, there is simply not enough evidence to decisively answer either of these questions.

Second, the remarks on ethics are often so dense that they need to be spelled out in order to become genuinely understandable, especially if they are to be of any relevance for contemporary philosophy. That is a well-known problem for scholars dealing with other aspects of Wittgenstein's work. But while we have thousands of pages of manuscripts and typescripts to consult when it comes to spelling out Wittgenstein's views on, say, language and mathematics, there is nothing of that sort

[3] When a reference is to a numbered remark, this is indicated by '§'; otherwise reference is to a page.

[4] These explanation are discussed in Barrett (1991: Chapter 12).

when it comes to his views on ethics, except from a few drafts of 'A Lecture on Ethics' (published in Zamuner et al. 2014) and the *Nachlass* remarks mentioned earlier.

Third, Wittgenstein's remarks on ethics occur in philosophical works as well as in personal documents. Should we treat the 'philosophical' remarks as more important or authoritative than the 'personal' ones? What is the status of remarks made in lectures and conversations (or rather, remarks which people report to have been made by Wittgenstein in lectures and conversations)? Although these questions can be asked with regard to other aspects of Wittgenstein's work as well, they are particularly relevant when it comes to ethics for several reasons. Because the philosophical works do not seem to offer enough material to enable us to work out Wittgenstein's ethical views, supplementation with personal remarks becomes necessary rather than optional. Moreover, Wittgenstein emphasizes that ethics is personal (Wittgenstein 2003: 85). Does that mean that the role and importance that we should assign to personal remarks is different in ethics than elsewhere?

These difficulties have not prevented scholars from trying to spell out Wittgenstein's ethical views. Because his explicit remarks are scarce, there is a lot of room for interpretation and there are many gaps to be filled with what is thought to be implicit in Wittgenstein. As a result, accounts of Wittgenstein's ethical views are highly divergent, and Wittgenstein has been associated with almost the entire spectrum of –isms in contemporary ethical theory (although, ironically, no one disputes that he had no theoretical aspirations). The only common ground between commentators seems to be that he eschewed reductionism and scientism in moral philosophy as elsewhere and that he did not offer normative guidance in the form of a decision procedure that could help us to solve moral problems or to decide what to do in particular cases.

The scarcity of Wittgenstein's remarks on ethics constitutes a major problem for those who want to spell out Wittgenstein's ethical views. I do not claim that these views cannot be spelled out or integrated into a coherent account. Rather, I think that, because of the scanty material that we have, any such account will necessarily either have to lean heavily on remarks with doubtful authority or to go *beyond* Wittgenstein's explicit views on ethics. It will then be an account inspired by or on the basis of Wittgenstein's ethical views.

1.1.2 How to Get Clearer on Wittgenstein's Ethical Views

Wittgenstein's remarks on ethics are too short and scattered to constitute, in themselves, an important contribution to moral philosophy. In order for them to become philosophically interesting, they need to be supplemented by something that enables us to interpret and understand them. The most obvious candidates for supplementation are to be found in the rest of Wittgenstein's philosophy.

How can Wittgenstein's philosophy, that is, everything in it that does not deal explicitly with ethics, help us to understand his ethical views? Wittgenstein frequently stresses the similarities between ethics and aesthetics (Wittgenstein 2005a: §6.421) and those between ethics and religion (Wittgenstein 1998: 5; Wittgenstein 1979b: 118). Wittgenstein's remarks about aesthetics and religion may, therefore, tell us something about his ethical views. It is doubtful, however, whether these remarks will offer more than an occasional insight. Readers of Wittgenstein have often been tempted to assimilate two things between which Wittgenstein stresses certain similarities, while tending to overlook the differences, even though Wittgenstein often points these out (see, for example, Chap. 2 on the similarities and differences between philosophical methods and therapies; for a difference between aesthetics and ethics, see Wittgenstein 1979a: 83). We cannot just transfer what Wittgenstein says about aesthetics or religion to ethics, and whether in particular cases we can or cannot do so will often be a matter of dispute. The main reason for that is well-known: just like the remarks on ethics, those on aesthetics and religion are often short, scarce and scattered. Sustained treatments of these topics are difficult to find in Wittgenstein's work, the remarks often occur in personal documents rather than in philosophical work, and when they do occur in philosophical work, it is often in lecture notes (Wittgenstein 1966, for example) and not in Wittgenstein's extensively revised, most authoritative works such as the *Tractatus* and the *Philosophical Investigations*.

Another way of appealing to Wittgenstein's philosophy in order to shed light on his explicitly ethical remarks seems more promising. Instead of looking at Wittgenstein's views on particular topics that he explicitly related to ethics, one can assess the relevance of other aspects of his work for moral philosophy. One might think here, first, about the usefulness of introducing some of Wittgenstein's terminology into the moral discourse. Is it helpful for moral philosophers to speak, for example, of moral language-games and forms of life, moral aspect-seeing and moral blindness, moral certainty, family resemblance concepts, etc.? Second, one could look into the consequences of some of Wittgenstein's arguments for moral philosophy: the private language argument, for example, or the remarks about rule-following and their relevance for our understanding of what it means to follow moral rules. Third, some of the things that Wittgenstein emphasizes again and again also seem to merit some emphasis in moral philosophy: the role of training and learning, the embeddedness of speaking and acting in concrete situations, the fact that justification comes to an end, and the influence of contingent general facts of nature on our ways of thinking and behaving.

In general, the idea here is to use Wittgenstein's philosophy as a kind of toolbox. Instead of just looking at how Wittgenstein deals with certain philosophical problems and trying to get clearer on his particular (dis)solutions of them (which is without doubt an interesting and useful task), I will take the liberty to use Wittgensteinian tools in order to tackle problems that Wittgenstein himself did not or not substantively deal with. Wittgenstein's philosophy is regarded not (or not only) as a set of views, but rather as offering *methods*. For our purposes, it suffices to say that philosophical methods are ways to approach, deal with or think about philosophical problems. The emphasis on methods rather than views is meant to stress that methods are what we *use* to approach problems. It is, of course, difficult and not always possible to separate views from methods. Take, for example, the famous dictum that 'the meaning of a word is its use in the language' (Wittgenstein 2009a: §43). Is this a view, competing with other views, such as the view that the meaning of a word is something mental, something in our heads? Or is it a methodological remark, telling us that if we want to know what a word means, we should look at how it is being used? In my view, it is both, but it is the methodological aspect that will interest me in what follows. As a view, it is a view in the philosophy of language and not in moral philosophy. As a method, it may well be used to approach problems in moral philosophy (such as, for example, the question 'What is virtue?').

Thus, the scope of the remark about meaning and use is certainly not to be confined to the philosophy of language. Wittgenstein did not see himself as an expert in one or several of the subfields of philosophy, as a logician or a philosopher of language, but simply as a philosopher. When he was asked under what title his courses at the University of Cambridge should be announced, he replied: 'The subject of the lectures would be philosophy. What else can be the title of the lectures but Philosophy' (Wittgenstein 2003: 340). Many of Wittgenstein's remarks, even when they explicitly concern topics in the philosophy of logic, language, mind or mathematics, can be read in a methodological light, as remarks in which Wittgenstein shows how to deal with philosophical problems in general. Moreover, a great deal of remarks is explicitly methodological or about philosophy in general. What is philosophy? How does it differ from science? How do its methods differ from those of science? How do we have to approach philosophical questions? What can we possibly hope to achieve in philosophy? These are questions that keep coming up throughout Wittgenstein's writings, and they occupy an important part of his works. If Wittgenstein has much to say about how to deal with philosophical problems in general, then why not use his thought for dealing with problems in *moral* philosophy?

1.2 Wittgenstein's Ethical Views and His Later Philosophy

The suggestion is that one can make up for the scarcity of Wittgenstein's explicit remarks on ethics by using the rest of his philosophy as a sort of toolbox with the help of which it becomes possible to interpret and understand these remarks better. Promising as this may sound, there are several problems with this suggestion. The major problem is that Wittgenstein's methods as well as many of his views have changed considerably throughout his philosophical career. Introductory philosophy textbooks traditionally distinguish between the first Wittgenstein of the *Tractatus* with his picture theory of meaning and search for the general form of the proposition, and the second Wittgenstein of the *Philosophical Investigations* with his conception of meaning as use, his language-games, etc. This view has come under attack in Wittgenstein scholarship in several ways. First, some scholars, especially those associated with *The New Wittgenstein* (Crary and Read 2000) and their so-called 'resolute' or 'therapeutic' reading of Wittgenstein, have argued that the picture of the two Wittgensteins obscures the continuities in Wittgenstein's thought.[5] Second, it has been argued that there are more than two Wittgensteins. According to Danièle Moyal-Sharrock (2004), Wittgenstein's post-*Investigations* work establishes the existence of a third Wittgenstein. Others contend that Wittgenstein's work between 1929 and 1937 is importantly different from both the *Tractatus* and the *Investigations* and claim that there is a middle Wittgenstein (Stern 1991).

It is not to be denied that there are important points of continuity between the *Tractatus* and the *Investigations*. The Tractarian idea that philosophy is not a theory but an activity (Wittgenstein 2005a: §4.112) and the contrast between philosophy and natural science (Wittgenstein 2005a: §4.1122) are present in the *Investigations* as well. At the same time, the differences between the *Tractatus* and the *Investigations* are conspicuous and should not be downplayed. Wittgenstein recognized 'grave mistakes' in what he set out in the *Tractatus* (Wittgenstein 2009a: 4). Compare, moreover, the hierarchical structure of the *Tractatus* and its decided style, which seems to leave no room for doubt about the 'unassailable and definitive' truth of its thoughts (Wittgenstein 2005a: 29) with the numerous questions, dialogues and tentative answers, 'criss-cross in every direction over a wide field of thought' (Wittgenstein 2009a: 3), in the *Investigations*.

Whether the differences between early and later Wittgenstein(s) warrant our speaking of one or two or more Wittgensteins will not be substantively discussed here. Maybe the issue is a matter of emphasis or philosophical temperament, and I will just note that Wittgenstein's own temperament seems to speak in favour of emphasizing differences rather than similarities. In his biography of Wittgenstein, Ray Monk refers to a conversation between Wittgenstein and his friend Maurice Drury:

> 'Hegel seems to me to be always wanting to say that things which look different are really the same', Wittgenstein told him. 'Whereas my interest is in showing that things which look the same are really different.' He [Wittgenstein] was thinking of using as a motto for his book the Earl of Kent's phrase form *King Lear* (Act I, scene iv): 'I'll teach you differences.' (Monk 1991: 536–537)

Regardless of how many Wittgensteins there are, a proper appreciation of the differences in different phases of Wittgenstein's thought requires me to be more specific about the methods or tools that I will use in order to gain a better understanding of Wittgenstein's ethical remarks. The most straightforward approach may be a chronological one. Suppose, for example, that I want to interpret a remark from 1945. Is it not obvious that, in using Wittgenstein's methods to do so, I should use his methods as they were in 1945? The problem is that, while Wittgenstein's methods clearly and substantially changed, I have not pronounced on whether his ethical views also changed. In fact, I have argued that, given the scarcity of the available material, that question has to be left open. If his ethical views did not substantially change, then a remark from 1945 could just as well have been made in 1917, which

[5] For a good overview of the authors and claims associated with the resolute reading, see Bronzo (2012).

raises the question why it would then have to be interpreted with the help of Wittgenstein's later methods.

The reasons for focusing on Wittgenstein's *later* philosophy are the following. First, a Tractarian approach to Wittgenstein's remarks on ethics seems either impossible or dogmatic. According to the *Tractatus,*

> The right method of philosophy would be this. To say nothing except what can be said, i.e. the propositions of natural science, i.e. something that has nothing to do with philosophy: and then always, when someone else wished to say something metaphysical, to demonstrate to him that he had given no meaning to certain signs in his propositions. This method would be unsatisfying to the other – he would not have the feeling that we were teaching him philosophy – but it would be the only strictly correct method. (Wittgenstein 2005a: §6.53)

How could this method help to clarify Wittgenstein's ethical views? And can it be practiced at all? According to the *Tractatus,* the philosopher's task is to provide the 'one and only complete analysis of the proposition' (Wittgenstein 2005a: §3.25), an analysis that breaks up complex propositions into elementary ones consisting of names (Wittgenstein 2005a: §4.221). But Wittgenstein does not give examples of such an analysis, and it remains unclear what the Tractarian 'names' are supposed to be. Moreover, there is the Tractarian view that 'there can be no ethical propositions' (Wittgenstein 2005a: §6.42) and that, consequently, 'ethics cannot be expressed' (Wittgenstein 2005a: §6.421). Ethics is one of the things whereof one cannot speak and whereof one must thus be silent (Wittgenstein 2005a: §7). In short, it seems as if a Tractarian approach to Wittgenstein's explicit ethical views (even to the ones 'expressed' in the *Tractatus* itself) is impossible.

It could be argued that the above reading of what a Tractarian approach involves is too strict. Perhaps we can disregard some Tractarian views on ethics and treat what Wittgenstein says about ethics as a set of propositions to be analysed in a Tractarian fashion. Perhaps problems with the interpretation of Tractarian names can be avoided if the method is described, for example, as one of substituting confusing modes of expression for other, less confusing ones (see Kuusela 2011). This method can be practiced, and examples can be found in the *Tractatus.* Take, for example, Wittgenstein's invention of the T/F notation (Wittgenstein 2005a: §4.31), 'which demonstrates the dispensability of the logical connectives in the propositional calculus, and rids one of the temptation to construe them as names of logical entities (e.g. of functions)' (Baker and Hacker 2005: 293). Consider his proposed dissolution of Russell's paradox in §3.333, or the idea of using different signs for 'is' as a copula (as in 'the rose is red') and 'is' as an identity sign (as in 'two plus two is four') in §3.323. When understood in this way, the method of the *Tractatus* can be practiced, and it does not seem impossible to apply it to ethical statements: could it not be illuminating to substitute confusing modes of ethical expression for less confusing ones? Certainly, but this method, although it is practiced in the *Tractatus,* is not absent from the *Investigations*; it is even explicitly advocated there (Wittgenstein 2009a: §90, §558). It is not exclusively Tractarian. The problem with the *Tractatus* is not that expressions cannot be clarified by substituting confusing modes of expression for others, but that Wittgenstein was looking in the *Tractatus* for a particular kind of notation, an ideal notation that would help to dissolve *all* philosophical problems. According to Oskari Kuusela, the problem is that Wittgenstein 'credited a certain logical model with the status of *the* ultimate means of clarification' (Kuusela 2011: 613). It is not *wrong* to substitute confusing modes of expression for others. Rather, it is dogmatic to believe that one particular mode of expression or notation will help to clarify all philosophical confusions.

Thus, if a Tractarian approach to Wittgenstein's views on ethics can be described in a way that does not make it impossible to practice, it still carries a dogmatic presupposition with it. Without this presupposition, the idea that confusing statements can be clarified by substituting one mode of expression for another loses its distinctively Tractarian character and can perfectly well be included in an approach based on Wittgenstein's *later* methods. Gordon Baker and P.M.S. Hacker mention the different notations for 'is' as copula and 'is' as an identity statement and the T/F notation as exemplifying Wittgenstein's *later* methods (Baker and Hacker 2005: 293; see, in this respect, also Chap. 3, footnote 6).

The first reason for taking Wittgenstein's later methods instead of his earlier ones is thus a negative one: a Tractarian approach seems either impossible or dogmatic. Things are different for an approach based on Wittgenstein's later philosophy. In the later works there is not one strictly correct method, but a whole series of methods, and the later works are full of examples in which these methods are practiced. In other words, while the Tractarian toolbox can be said to contain only one tool, the later toolbox contains a great variety of tools, including a version of the Tractarian tool. It is therefore much more likely to contain tools with the help of which Wittgenstein's remarks on ethics can be clarified. Second, although Wittgenstein does not apply his methods to problems in moral philosophy in his later works, these works do not contain an explicit or implicit rejection of the possibility of doing so. On the contrary, they sometimes seem to endorse the possibility of using Wittgensteinian methods in moral philosophy (Wittgenstein 1966: 1, 1980a: §160; see also Chap. 3, Sect. 3.5). Third, the later works are not hierarchically structured and not in the business of building a system, as is the *Tractatus*. The remarks stand next to each other on an equal hierarchical level. Consequently, the later works lend themselves better to using one remark or another without having to take a whole system with it.

I should specify my use of 'later' in 'Wittgenstein's later methods'. By 'later methods' or 'Wittgensteinian methods' I mean primarily the methods that Wittgenstein explicitly discusses in the *Philosophical Investigations* (see especially Wittgenstein 2009a: §§89–133). That does not mean that references to the *Tractatus*, to the middle period or to the post-*Investigations* works are necessarily out of place. As we have seen, there are important continuities between the *Tractatus* and the *Investigations*, not to speak of continuities between the *Investigations* and the middle period or between the *Investigations* and the post-*Investigations* period. Thus, it often happens that Wittgenstein says something in pre- or post-*Investigations* works that is compatible with, and could as well have been said in, the *Investigations*. In such cases, references to those other works may well serve to illustrate or clarify Wittgenstein's position in the *Investigations*, and the use of materials from other works will then count as 'using Wittgenstein's later methods'. Wittgenstein's later methods, as I will understand them here, are methods used in the *Investigations* and methods in line with those used in the *Investigations.*

Using the later Wittgenstein's philosophical methods in order to shed light on Wittgenstein's own ethical remarks is something that *can* be done. The question then is: why *should* we do it? My answer is that doing so will enhance our understanding of some of Wittgenstein's ethical views, but I will not attempt to support that claim in a general way. Rather, I will provide and work out examples of how Wittgenstein's later methods can help us to get clearer on particular remarks about ethics that Wittgenstein made (Chaps. 4 and 5). To be sure, I do not claim and I do not think that Wittgenstein's later methods will enable us to get clear on every remark about ethics that he made. My claim is more modest: some remarks will be better understood and found to be of philosophical interest when placed in the context of Wittgenstein's later methods.

1.3 Wittgenstein's Later Philosophy and Contemporary Moral Philosophy

I began this chapter by asking why Wittgenstein's thought would be relevant for moral philosophy. After having considered some problems with the idea that his contribution lies in his explicit remarks on ethics, I suggested that Wittgenstein's later methods could help to understand at least some of his ethical views in a philosophically interesting way. But if these later methods can be rewardingly applied to the interpretation of Wittgenstein's own ethical views, then why not apply them to problems in contemporary moral philosophy that have no obvious link with these views? Why not ask: how can problems in moral philosophy best be approached in the light of Wittgenstein's later philosophy? The best approach will then be understood as the one that, using Wittgensteinian tools, best succeeds in contributing to debates in contemporary moral philosophy, the one that is philosophically (rather than exegetically) more successful.

The uncoupling of Wittgenstein's later methods from his ethical views avoids several problems with these views. First, in contrast to his ethical views, there is no lack of material on Wittgenstein's later methods in his philosophical works.

Second, Wittgenstein's ethical views have not only been explained in very different ways, these views are also very differently assessed. While some think that we have much to learn from Wittgenstein's remarks on ethics, others have been irritated by religious and mystical elements in them or accuse Wittgenstein of aestheticizing ethics. While Wittgenstein scholars are often quick to stress how important his remarks on a certain topic are, they do not agree at all in their evaluation of his ethical views. Hans-Johann Glock points out that Wittgenstein expressed anti-Semitic views during the 1920s and 1930s and that he was prone to misogynist sentiments. He concludes that 'as a philosophical dialectician, Wittgenstein was a genius. But as a moral mentor and cultural critic he was a loose cannon' (Glock 2015: 100).[6] Glock might have misinterpreted some of Wittgenstein's remarks (for a worked-out example, see De Mesel 2015), but it remains a fact that some of Wittgenstein's ethical views were, as Duncan Richter puts it euphemistically, 'quite unorthodox' (Richter 2002: 325), and that some of them seem utterly unreasonable (for examples, see Richter 2002). If Wittgenstein's later methods can effectively be applied to problems in moral philosophy that have no obvious link with Wittgenstein's (questionable) ethical views, his contribution to moral philosophy need no longer be regarded as intimately tied to these views.

Third, it is at least doubtful whether Wittgenstein regarded his explicit remarks on ethics as a contribution to moral philosophy. He repeatedly suggested that the very enterprise of doing moral philosophy is deeply problematic, undesirable or impossible. Such suggestions are prominent in the *Tractatus* and in 'A Lecture on Ethics'. Although they occur less in later writings in which, as we have seen, the possibility of applying his philosophical methods to problems in moral philosophy is not excluded and even sometimes explicitly hinted at, doubts about the very possibility of doing Wittgensteinian moral philosophy cannot be entirely put aside. Especially for those who see no significant changes in Wittgenstein's ethical outlook after the *Tractatus,* it will remain problematic to take his ethical views as a starting point for doing moral philosophy and to apply his philosophical methods to problems that he might not have considered as philosophical problems at all.

One could object here that Wittgenstein had doubts not only about the possibility of doing *moral* philosophy but about the possibility of philosophy in general. Did he not think that philosophy *as such* was in trouble? There is a crucial difference between the two cases. While he certainly had serious reservations about the way in which philosophy was practiced (and would have serious reservations about the way in which it is still practiced), he did not doubt the possibility of doing philosophy. According to Wittgenstein, there is a need for philosophy, only for a different kind of philosophy than the philosophy that we are used to: 'One might say that the subject we are dealing with is one of the heirs of the subject which used to be called "philosophy"' (Wittgenstein 1969: 28). Things are different for moral philosophy. In a conversation with members of the Vienna Circle recorded by Friedrich Waismann, Wittgenstein says: 'At the end of my lecture on ethics, I spoke in the first person. I believe that is quite essential. Here nothing more can be established, I can only appear as a person speaking for oneself' (Waismann 1965: 16). If we can speak about ethics at all, this speaking will be speaking for oneself, and not some kind of philosophy. According to Wittgenstein, 'what is ethical cannot be taught' (Wittgenstein 1979b: 117). He condemned all the 'claptrap about ethics', and it is obvious that he meant *philosophical* claptrap, not merely what ordinary people tend to say about ethics: the claptrap concerns such questions as 'whether intuitive knowledge exists, whether values exist, whether the good is definable' (Wittgenstein 1979b: 68–69). In concentrating on later methods rather than views, Wittgenstein's

[6] Glock refers to some more charitable assessments of Wittgenstein's ethical views: see DeAngelis (2007); Klagge (2011); Lurie (2012); Szabados (2010).

views about the (im)possibility of moral philosophy do not have to trouble us as much as they would if we were confining ourselves to clarifying his ethical views, with or without the help of his methods.

Several problems with Wittgenstein's ethical views can thus be avoided by focusing on his philosophical methods rather than on his ethical views. But obviously there should be more to say in favour of applying Wittgenstein's later methods to debates in moral philosophy than that it does not have certain problems. What can we hope to achieve by doing that? Why expect it to be philosophically interesting? In general, my hypothesis is that Wittgensteinian approaches to problems in moral philosophy may yield results analogous (and of comparable philosophical interest) to those obtained by applying his later methods to problems in other branches of philosophy. What are these results?

Roughly speaking, the overarching aim of Wittgenstein's later philosophical methods lies not in providing answers to philosophical questions, but in showing that there is something wrong with these questions so that there *is* no answer to them. Questions asked by philosophers often fail to make sense or cannot be meaningfully answered, and I will give three examples of (related) ways in which this may be the case. First, questions can fail to make sense because it has not been established what use of words or expressions is in question. Take, for example, the questions 'Is this object composite?' and 'What are the simple constituent parts of a chair?' (Wittgenstein 2009a: §47). These questions, Wittgenstein says, cannot be answered unless it is established what uses of 'simple' and 'composite' are in question.

> Asking 'Is this object composite?' *outside* a particular game is like what a boy once did when he had to say whether the verbs in certain sentences were in the active or passive voice, and who racked his brains over the question whether the verb 'to sleep', for example, meant something active or passive. (Wittgenstein 2009a: §47)

Wittgenstein emphasizes that 'we use the word "composite" (and therefore the word "simple") in an enormous number of different and differently related ways' (Wittgenstein 2009a: §47). The particular context in which words and expressions are used often gives us a clue about how to understand them. However, according to Wittgenstein, philosophers typically take words and expressions out of the contexts in which they are 'at home' (Wittgenstein 2009a: §116). As a consequence, it becomes unclear how philosophical questions are to be understood: 'To the *philosophical* question "Is the visual image of this tree composite, and what are its constituent parts?" the correct answer is: "That depends on what you understand by 'composite'." (And that, of course, is not an answer to, but a rejection of, the question.)' (Wittgenstein 2009a: §47). A Wittgensteinian approach to philosophical questions often involves showing that these questions have not been meaningfully formulated, because it has not been established what uses of words or expressions are in question.

Second, in some cases it cannot even be established what uses of words or expressions are in question because words and expressions are uttered in ways and contexts in which they *cannot* be meaningfully used, although it seems as if they are being so used. An example is the question 'Where does the present go when it becomes past?' (Wittgenstein 1969: 107). According to Wittgenstein, philosophers are often misled by certain analogies between the forms of expression in different regions of our language. We say that moving objects pass by and that time passes by. This analogy makes us think that, because we can meaningfully ask where a moving object has gone, we can meaningfully ask where time has gone. However, to ask this is to overlook disanalogies between time and objects: while objects can be located in time and space, time itself has no spatial location.

A third way in which philosophical questions cannot be meaningfully answered is when they involve an unwarranted assumption. Consider, for example, the question 'What do all games have in common in virtue of which we call them "games"?' The answer to this question is not that games have this or that in common, but that we need not assume that games have anything in common in virtue of which they are called 'games' (Wittgenstein 2009a: §66). Some concepts are family resemblance concepts (Wittgenstein 2009a: §67). The idea that all instances of a concept *must* have something in common in virtue of which they fall under that concept is nothing more than an unwarranted assumption, a picture that holds philosophers captive (Wittgenstein 2009a: §115).

In this case too, analogies between forms of expression in different regions of our language have misled philosophers. An example is the question 'What kind of object is pain?' We are led to think that pain is a kind of object because we can say both of objects and of pain that we have them and that we perceive them. Both objects and pain can be located: I can ask where your pen is and where your pain is, etc. However, the fact that 'pain' is sometimes used in ways analogous to the use of object-words does not mean that pain *is* an object. Instead of taking sides in a debate about what kind of object pain is, a Wittgensteinian approach will consist in showing that all parties in the debate share the unwarranted assumption that pain is an object.

Why do questions asked by philosophers fail to make sense? In everyday life, words and expressions are used in contexts which give them sense: they are embedded in language-games, in what we do, in our practices, in our form(s) of life (Wittgenstein 2009a: §23). When philosophers take words and expressions out of these contexts, they take them out of the environment that gives them their sense, in which they make contact with our lives. Wittgenstein describes the situation as follows: 'We have got on to slippery ice where there is no friction, and so, in a certain sense, the conditions are ideal; but also, just because of that, we are unable to walk. We want to walk: so we need *friction.* Back to the rough ground!' (Wittgenstein 2009a: §107). The situation is comparable to one in which one takes an animal out of its natural environment into a vacuum where it cannot live in order to study its behaviour, while its behaviour will only be understood when it is observed in its natural environment. That is why Wittgenstein says that philosophers have to 'bring words back from their metaphysical to their everyday use' (Wittgenstein 2009a: §116). The philosopher's task is to show that philosophical questions make no sense because words and expressions have been used out of the (linguistic and non-linguistic) contexts in which they can play a meaningful role.

In summary, the application of Wittgenstein's later methods to philosophical questions often results in showing that these questions make no sense and therefore cannot be meaningfully answered. How can these results be achieved? This is a question about Wittgenstein's later methods, and these will be discussed in more detail in Chaps. 2 and 3. A prominent way to achieve these results is to emphasize the multiplicity of uses of words and expressions by providing a 'surveyable representation' (Wittgenstein 2009a: §122), a kind of overview, of their use. An overview of the use of 'composite' will display the 'enormous number of different and differently related ways' in which 'composite' can be used. An overview of the use of 'game' will make clear that we do not call games 'games' on the basis of a common characteristic. An overview of the uses of 'time' and 'pain' will establish that, although these words are often used in ways analogous to object-words, they also have uses that object-words do not have. In his later works, Wittgenstein describes the use of a great variety of words and expressions. In the *Investigations,* there are discussions of, for instance, uses of 'understanding', 'sensation', 'thinking', 'imagining', 'consciousness', 'intention', 'meaning', 'language', 'expectation', 'experience', 'rule', 'belief', 'hope', 'feeling', 'willing', 'remembering' and 'seeing'. These discussions contain examples of actual word-use, fictitious examples and invented language-games in which it is shown how words and expressions can and cannot be meaningfully used, and numerous comparisons that highlight differences and similarities between uses of words and expressions (sometimes across different regions of our language).

I believe that, in general, Wittgenstein has been successful in showing, by way of his later methods, that certain philosophical questions in several subfields of philosophy (philosophy of mind, philosophy of psychology, philosophy of logic, philosophy of language, philosophy of mathematics) make no sense. Why, then, should we not expect analogous results from the application of his methods to problems in moral philosophy? There seem to be no principled reasons against the possibility of showing that some questions in moral philosophy cannot be meaningfully formulated or involve unwarranted assumptions. Moral philosophers too may well have been misled by analogies between the forms of expression in different regions of our language. Moral expressions are deeply embedded in our lives and interwoven with our practices, and to study them in isolation from their role in our lives and practices will probably yield distorted results. Why would certain concepts in moral philosophy not be family resemblance concepts (Wittgenstein 2009a: §77), so that it will be misguided to ask what all their instances have in common? Words and expressions in moral philosophy are used in a multiplicity of ways, and overviews of these ways by way of comparisons, examples of actual word-use and invented language-games may make widespread confusions or ambiguities disappear.

These are the kind of results that can be expected from applying Wittgenstein's later methods to problems in moral philosophy. In what follows, I will provide and work out examples of such applications to particular questions, problems and debates in contemporary moral philosophy (Chaps. 6, 7 and 8). Again, as in the previous section with regard to the application of Wittgenstein's methods to his own ethical views, I do not claim that Wittgenstein's later methods will enable us to develop significant contributions to every debate in moral philosophy. The claim is that some problems, questions, views and debates will be better understood when approached with the help of Wittgenstein's later methods.

1.4 Structure and Overview of the Book

The overall aim of this book is to show that Wittgenstein's later methods can fruitfully be applied to several problems in moral philosophy. I will now turn to an overview of how that claim will be defended. The book consists of three parts. In the first part, I discuss Wittgenstein's later philosophical methods. The second part is an investigation of how these methods help to understand some of Wittgenstein's ethical views. In the third part, Wittgenstein's later philosophy is applied to problems in contemporary moral philosophy. In the conclusion, I indicate how the topic(s) of the book can be developed in future research.

1.4.1 Part I. Wittgenstein's Later Philosophy

The main claim of the book concerns the applicability of Wittgenstein's later methods to problems in moral philosophy. The first question that suggests itself is: what are these later methods? Because I do not want to engage here in debates (however interesting they may be) about similarities and differences between several later works, I will understand by 'later methods' in the first place the ones that Wittgenstein sets out and applies in the *Investigations*. The first part of the book is devoted to the question of what these later methods are.

It is not my aim to give anything like a full account of Wittgenstein's later methods. Good accounts are available, and I have been influenced in particular by Baker and Hacker's impressive analytical commentary on the *Investigations* (Baker and Hacker 2005, 2009a, b; Hacker 1993a, b, 2000a, b), Kuusela (2008), McGinn (1997) and Schroeder (2006). Instead of summarizing the literature, I will approach the question of what Wittgenstein's later methods are by focusing on what I see as two key paragraphs in the *Investigations* (Wittgenstein 2009a): §122 and §133. These paragraphs serve as starting points from which to explain how Wittgenstein thought about philosophical problems and how these problems should be approached. Moreover, a focus on these paragraphs allows me to review and criticize some of the secondary literature on Wittgenstein's later philosophical methods from a specific angle. I thus intend to throw light on Wittgenstein's later methods in the form of original contributions to the secondary literature on the *Investigations.*

Let me start with §133. In that paragraph, Wittgenstein famously compares philosophical methods to therapies: 'There is not a single philosophical method, though there are indeed methods, different therapies, as it were.' The comparison of philosophical methods to therapies has touched a chord with many people: it has been amply discussed and interpreted in many, sometimes highly different ways. In Chap. 2, I argue that some of these ways cloud rather than shed light on Wittgenstein's philosophical methods.

My way of clarifying Wittgenstein's comparison of philosophical methods to therapies has affinities with what Dedre Gentner (1983) has called 'structure-mapping' in a seminal paper on analogy.[7] The idea is the following. Suppose that I compare the nucleus of an atom to the sun. Why is this comparison illuminating? Not because the nucleus and the sun have a number of important attributes in common. The nucleus is not yellow or hot or particularly big, while being yellow, hot and particularly big are three of the most striking attributes of the sun. The comparison is illuminating because it maps important relational features of the sun onto the nucleus: the sun stands to the planets of our solar system as the nucleus stands to the electrons. 'Sun', 'solar system' and 'planet' are interrelated concepts, and the one is often explained by using the other. Insight into the comparison between the sun and the nucleus can be gained by asking: what stands to the nucleus as the planets stand to the sun? Or: if the nucleus corresponds to the sun, what corresponds to the planets and the solar system?

The concept 'therapy' seems to be primarily related to those of a patient, an illness or a disorder, a therapist and an ideal of health. The idea is that we can gain insight into the comparison between therapies and philosophical methods by focusing on these relational features: if philosophical methods correspond to therapies, what corresponds to the illness or disorder, the patient, the therapist and the ideal of health in philosophy?

An important conclusion of Chap. 2 is that Wittgenstein's *comparison* between philosophical methods and therapies has often been read as an identity statement: philosophical methods *are,* literally, a kind of therapy and since philosophers need them, they must be *literally* ill. Philosophical methods are not candidates for some kind of physical therapy, so they must be some kind of psychological therapy. These interpretations are not uncommon, but I believe that they are deeply mistaken and misleading. Moreover, showing them to be false has a particular importance for my overall project. If philosophical methods were psychological therapies, directed at taking away people's 'mental disturbances' (Read and Hutchinson 2010: 153) or 'individuals' *troubled states of mind*' or their internal conflicts (Baker 2004b: 212), then I would not have undertaken this project, because apart from the exegetical problems that I have with the idea as an interpretation of Wittgenstein, I also have a fundamental substantive problem with it: I disagree with the whole idea and cannot see what would be gained philosophically by treating problems in moral philosophy as mental disturbances. Not only would I not have undertaken the project if the psychological reading were true, I also could not have hoped to be successful. I do not think that I am able, through my philosophical work, to make mental disturbances disappear or to take people's internal conflicts away.

Although Chap. 2 has something to say about Wittgenstein's philosophical methods and his conception of philosophy, its thrust is mainly negative: certain misleading

[7] I would like to thank Lorenz Demey for drawing my attention to Gentner's paper.

misinterpretations are being scrutinized and the contribution of the article lies primarily in its claims about what Wittgenstein's philosophical problems and methods are *not:* philosophical problems are not illnesses, philosophical methods are not psychological therapies and they are not ultimately aimed at the end of philosophy. The discussion of §122 in Chap. 3 is, in general, more positive, as it seeks to offer a novel interpretation of Wittgenstein's concept 'surveyable representation' (*übersichtliche Darstellung*), of which he explicitly says that it is 'of fundamental significance for us' (Wittgenstein 2009a: §122). Although it too shows two dominant interpretations of Wittgenstein's notion of a surveyable representation to be inadequate, these interpretations are not wholly rejected but rather reconciled in my own interpretation.

I should say something here about the relation between §122 and §133. It is often suggested that these two crucial paragraphs put forward incompatible and even contrasting ways of doing philosophy. While, according to §122, the philosopher should do something positive or constructive, namely offer a surveyable representation of the use of our words, §133 seems to regard the philosopher's task as negative: (s)he should make philosophical problems disappear. A comparison between other passages in Wittgenstein's later philosophical works seems to result in the same contrast. On the one hand, the task of the philosopher is described as 'arranging the books of a library': the books belonging together are being put together, the books not belonging together are being put on different shelves (Wittgenstein 1969: 44). On the other hand, the philosopher is someone who destroys houses of cards (Wittgenstein 2009a: §118). I do not think that these two tasks, when rightly understood, stand in tension with each other. The therapeutic task of philosophy is to get us to see that the philosophical problems with which we are struggling are based on conceptual confusions. Once we see that, the problem disappears (negative). Providing surveyable representations (positive) is a way of getting us to see that, a method of therapeutic philosophy. The positive and negative tasks of philosophy are inextricably intertwined.[8] This is well illustrated by yet another comparison in which Wittgenstein sees doing philosophy as 'tidying up a room' (Wittgenstein 1980c: 24; see also Chap. 2). On the one hand, this is a negative task: all the rubbish is being thrown away. On the other hand, the room will not be tidy until we have put everything in its place.

Apart from offering an interpretation of the concept 'surveyable representation', Chap. 3 also discusses the possibility of applying Wittgenstein's later methods to problems in moral philosophy and thus paves the way for the second and third part of the book. First, Wittgenstein's later methods are confronted with his own work on ethics, in particular with 'A Lecture on Ethics'. As a result, a received view about the interpretation of the lecture, concerning the place of the lecture in Wittgenstein's work, is found to be one-sided. I argue that, instead of seeing the lecture as belonging to the work of the early, Tractarian Wittgenstein, as almost all scholars do, it is more accurate to regard it as a transitional text in Wittgenstein's oeuvre.

[8] I agree on this point with Baker (2004a: 36). A contrast or tension between the positive and negative tasks of the Wittgensteinian philosopher is suggested, for instance, by Hacker (1996: 111, 2001: 333, 338) and Kenny (1982: 38–39, 42–43, 2004: 175).

Second, Wittgenstein's later methods are applied to problems in moral philosophy that he did not himself discuss. The problem that I introduce in Chap. 3 concerns the semantic function of moral judgments. On one side of the debate, cognitivists claim that moral judgments express beliefs, while on the other side non-cognitivists claim that they express 'non-beliefs' such as emotions or prescriptions. I defend the idea that giving a surveyable representation of how moral judgments are used may unearth an assumption common to both parties in the debate, namely that all moral judgments must have a semantic function in common.

1.4.2 Part II. Wittgenstein's Later Philosophy and His Ethical Views

The second part of this book constitutes a detailed exploration of the approach outlined in Sect. 1.2. The overall claim is that some of Wittgenstein's ethical views will be better understood when read in the light of his later philosophical thought. The problem is not that Wittgenstein scholars have denied this: Wittgenstein's later methods are often explicitly mentioned in interpretations of his ethical views. Rather, the problem is that scholars have not realized the full consequences of their appeal to Wittgenstein's later methods and, as a result, have presented a distorted understanding of Wittgenstein's ethical views.

Chapter 4 provides an example. In that chapter I discuss four interrelated claims that have been made in strikingly similar ways by several Wittgenstein scholars. One of these claims is that 'moral' is a classification by use and not by subject matter. This claim has been ascribed to Wittgenstein and defended partly by an appeal to his later philosophical methods. In arguing for it, its defenders lean heavily on the idea that the meaning of a word is its use in the language (Wittgenstein 2009a: §43). They thereby recognize the relevance of Wittgenstein's later philosophical thought for the interpretation of his ethical views. However, in my view, a serious misunderstanding is involved in what they take the consequences of §43 to be. It is in line with §43 to say that 'moral' is a classification by use, but that does not mean that it cannot, at the same time, be a classification by subject matter. Cora Diamond refers to 'Wittgenstein's view that a particular sentence ["20 apples plus 30 apples is 50 apples"] might belong to pomology or might belong to mathematics, and *which* it belongs to depends not on what it is apparently about but on its use' (Diamond 1996: 237). What Diamond's example shows, in my view, is that we cannot decide on what a sentence is about or on what its subject matter is without looking at how it is used: subject matter is determined by use. Diamond, however, concludes that 'pomological' is not a classification by subject matter but a classification by use. She suggests that there is a conflict or opposition between use and subject matter and that something cannot be a classification by subject matter if it is a classification by use. However, that does not follow from §43.

Chapter 4 not only shows how Wittgenstein's methods have been misused in the interpretation of his ethical views, it also makes clear that, while the relevance of some later remarks such as §43 is recognized, several other relevant remarks have been overlooked, which gives the insistence on §43 a somewhat arbitrary character. I will explain that by looking at another claim discussed in Chap. 4, related to the one that 'moral' is not a classification by subject matter: the claim that ethics has no particular subject matter. Defenders of this claim rightly note that the subject matter of moral philosophy cannot be sharply demarcated: sometimes we cannot make out, for example, whether a problem is a moral one or not. It seems obvious, on the other hand, that there are many cases in which we can say that a problem is or is not a moral problem. In short, the region of the moral has boundaries, but these boundaries are blurred. Or in terms of concepts: the concept 'moral' has blurred boundaries. In §71 (Wittgenstein 2009a), Wittgenstein discusses precisely the idea of a 'concept with blurred edges'. He asks whether a blurred concept can be called a concept at all. He does not straightforwardly answer that question, but asks further questions: 'Is a photograph that is not sharp a picture of a person at all? Is it even always an advantage to replace a picture that is not sharp by one that is? Isn't one that isn't sharp often just what we need?' Baker and Hacker comment, adding a reference to Wittgenstein (Wittgenstein 2005b: 196), that by asking these further questions, Wittgenstein's intention is to ridicule the initial question as dogmatic: 'What if someone were to say: "An out-of-focus picture really isn't a picture at all?"' They conclude, rightly in my view, that according to Wittgenstein 'for certain purposes, a blurred picture may be exactly what one wants; so too, for many purposes, concepts without sharp boundaries are exactly what we want [...]' (Baker and Hacker 2009a: 161).

That Wittgensteinian conclusion seems to be the opposite of what some Wittgenstein scholars have concluded. From the fact that the region of the moral has blurred boundaries, they infer that therefore there is no such region. Because there is no such region, they say, ethics has no particular subject matter. Their use of the very word 'region' (Mulhall 2002: 304) is significant, because Wittgenstein uses the very same term (Wittgenstein 2009a: §71): 'Frege compares a concept to a region, and says that a region without clear boundaries can't be called a region at all. This presumably means that we can't do anything with it. – But is it senseless to say "Stay roughly here?"' Baker and Hacker comment, again rightly in my view: 'Wittgenstein denies that a roughly indicated region is no region. So too, a concept that is not defined by specification of characteristic marks *is* a concept for all that, and may be more useful for certain purposes than one that is sharply circumscribed' (Baker and Hacker 2009a: 161). So while §71 denies that a roughly indicated region is no region, the authors discussed in Chap. 4 appear to think that a roughly indicated region is indeed no region: according to them, because the region of the moral has blurred boundaries, there is no moral region.

The argument in Chap. 4 is not just an example of how Wittgenstein's later methods can enhance our understanding of his ethical views (or how misunderstanding these methods or arbitrarily selecting among them may cloud it), it also plays a crucial role in my overall project. Apart from the claims that (1) ethics has no particular subject matter and that (2) 'moral' is not a classification by subject matter, I take issue with the claims that, according to Wittgenstein, (3) there is and can be no such thing as meta-ethics and that (4) ethics is not a distinguishable region of life. I show that (3) and (4) are often taken to follow from (1) and (2), although they do not follow: even if 'moral' or 'ethical' would not be classifications by subject matter (because ethics is thought to have no subject matter) but only by use, meta-ethical reflection on the moral use would not thereby become impossible and the moral use could well be distinguishable from other (kinds of) uses. The importance of arguing against (3) and (4) for my project is that, if indeed there can be no such thing as meta-ethics, and if indeed ethics is not a distinguishable region of life, then the project of applying Wittgensteinian methods to problems in meta-ethics would become plainly impossible, since there would be no such things as 'problems in meta-ethics'.

In Chap. 5, I examine another case in which some of Wittgenstein's later methodological points have been overlooked. Mario Brandhorst has recently claimed that, while Wittgenstein leaves room for truth and falsity, facts, correspondence and reality in ethics, his target is objectivity. I accept the first part of Brandhorst's claim and the reasons and arguments that he invokes in support of it: Wittgenstein indeed leaves room for truth and falsity, facts, correspondence and reality in ethics. I argue, however, that Brandhorst's arguments in favour of the first part of his claim invite similar arguments in favour of objectivity (for Brandhorst's reply, see Brandhorst 2017).

The Wittgensteinian methodological points that Brandhorst uses are various. He recognizes that words such as 'fact' and 'reality' can be used in a variety of different ways in different language-games, that the meaning of a word is its use in the language, that talk of truth and reality may not have metaphysical depth, and so on. I contend that *all* the Wittgensteinian tools that Brandhorst uses can be used in favour of an argument *for* rather than against objectivity in ethics. That is, exactly the same kind of argument that led him to his first claim leads to serious doubts about his objectivity claim. Again, I discern a certain arbitrariness in the argument: Wittgenstein's later thought is invoked in support of one claim, but forgotten or misrepresented in the discussion of another.

The problems with the accounts that I criticize in Chaps. 4 and 5 are, in a sense, remarkably similar. In 'Questions and Answers in Wittgenstein's Philosophy', Oswald Hanfling discusses several ways of getting into difficulties with Wittgenstein. One of these, he says, is

> [...] a craving to look for hard and fast claims where what one is given is tentative investigations; for proofs and refutations where 'description alone' is intended; and for answers were only questions are given. This is not to imply that Wittgenstein did not have firm views, that he never gave any arguments, or that his philosophy is devoid of answers. But these features are not as prevalent as is sometimes assumed, and a danger lies in reading them into the text when this is not what was intended. There is a passage in which Wittgenstein referred explicitly to his preference for questions. 'In philosophy', he wrote in *Remarks on the Foundations of Mathematics*, 'it is always good to put a *question* instead of an answer to a question.' (He went on, however, to question this statement!) (Hanfling 2002: 12)

I believe that the authors criticized in Chaps. 4 and 5 have read refutations into Wittgenstein where there are none. They have been looking for answers where there are no hard and fast answers to be found. They claim that, according to Wittgenstein, there is *no* meta-ethics and *no* objectivity in ethics. It should be emphasized that I have not defended the opposite claims: I am not saying that, according to Wittgenstein, there *is* meta-ethics and objectivity in ethics. Rather, I believe that Wittgenstein in no way *excludes* these, that he *leaves room* for them, and that on the basis of Wittgenstein's work a sound argument against them cannot be developed. What I argue against in both Chaps. 4 and 5 is a form of dogmatism in reading Wittgenstein.[9]

1.4.3 Part III. Wittgenstein's Later Philosophy and Contemporary Moral Philosophy

In the third part of this book, I will be looking at contributions that Wittgenstein's later philosophy can make to contemporary moral philosophy. In Chap. 6, I outline a Wittgensteinian approach to the debate on moral perception. On one side of the debate are those who think that moral perception is not really a form of perception, because there are too many significant differences between moral perception (seeing wrongness, for example) and other, uncontroversial kinds of perception (seeing a tree, for example). On the other hand, there are those who defend the possibility of moral perception, often by invoking an analogy with colour perception: seeing moral value, they argue, should be compared to seeing red rather than to seeing a tree.

Both parties have made valuable points. Critics of moral perception have rightly pointed out that moral perception is significantly different from certain paradigm cases of perception and even from seeing colour. Defenders of moral perception have rightly claimed that if we can provide examples of non-moral perception that uncontroversially count as cases of perception and that are relevantly similar to cases of moral perception, the case for moral perception will be strengthened. I argue that seeing emotion is relevantly similar to seeing moral value and does not suffer from the defects of the colour analogy emphasized by critics of moral perception. Thus, if defenders of moral perception seek to draw support from an analogy, then seeing emotion will better protect them against criticisms than seeing colour will.

In what respects is my approach Wittgensteinian? First, examples of seeing emotion occur in the context of Wittgenstein's discussions of aspect perception (Wittgenstein 2009b: §§227–228, 1980b: §170, 1981: §225). Wittgenstein's remarks on aspect-seeing, as well as the secondary literature on that topic, have helped me to understand the specificity of seeing emotion and the ways in which it is linked to other kinds of seeing. Second, Wittgenstein often explicitly refers to colour-seeing in his discussions of aspect-seeing, mostly to point out important differences

[9] See, in that respect, the section on 'Diamond's Dogmatism' in Moyal-Sharrock (2012: 238–239). I came across Moyal-Sharrock's article after the publication of Chap. 4. There are important affinities between the two articles, and I believe that the fact that we have independently arrived at similar conclusions speaks in favour of them.

between them (Wittgenstein 2009b: §247 and §351). Third, I have not only been inspired by what could be called Wittgensteinian *views* on perception, but also by his *methods.*

(1) I believe that the colour analogy has become so dominant in the debate that participants on both sides of it have been held captive by the idea that, if there is such a thing as moral perception, it must be understood on the model of colour perception. Colour perception, however, is nothing more than what Wittgenstein calls an 'object of comparison' (Wittgenstein 2009a: §131). Even if one does not accept the conclusion that the emotion analogy is better than the colour analogy as an analogy to moral perception, it should at least be recognized that it offers an alternative to it. The fact that there are such alternatives should weaken the hold of the colour analogy on philosophers' thinking about moral perception, and I believe that trying to weaken the hold of certain analogies is a Wittgensteinian way of working ('by our method we try to counteract the misleading effect of certain analogies', Wittgenstein 1969: 7; see also Wittgenstein 1969: 28 and 49).
(2) Wittgenstein refers to 'the particular peace of mind that occurs when we can place other similar cases next to a case that we thought was unique' (Wittgenstein 2005b: 307). Philosophers have thought that the case of moral perception was unique, in the sense that it does not resemble other kinds of perception in ways significant enough for us to be justified in speaking of perception here at all. I have tried to show that the perception of emotion is similar to moral perception so that, when placed next to it, moral perception looks less unique and anomalous. Wittgenstein insisted that philosophy should produce 'that kind of understanding which consists in "seeing connections". Hence the importance of finding and inventing *intermediate links*' (Wittgenstein 2009a: §122). As I say in Chap. 6, I believe that seeing emotion is an intermediate link between seeing colour and seeing moral value.
(3) I have not argued for or against moral perception. Wittgenstein often does not take sides in debates, his methods are rather aimed at laying bare certain assumptions that all parties share. My point is not that moral perception is really a case of perception or that it really is not. On the one hand, there are reasons for defenders of moral perception to insist on the term 'perception' (instead of, say, 'judgment'). Jeremy Wisnewski and Henry Jacoby describe the case of a woman who, although she sincerely *judges* that one should not regard Jews as unclean, cannot help but *see* them as unclean. Her problem is not one of judgment or knowledge (Wisnewski and Jacoby 2007: 232). Moreover, 'perception' suggests directness and immediacy in a way that 'judgment' does not (Wisnewski and Jacoby 2007: 234–235). On the other hand, critics may argue that the requirement of concept-mastery is alien to genuine forms of perception (Wright 1988: 12). Wittgenstein summarizes as follows: '"But this surely isn't *seeing*!" – "But this surely *is* seeing!" – It must be possible to give both remarks a conceptual justification' (Wittgenstein 2009b: §181). In the end, whether moral perception will count as a genuine form of perception will crucially depend on

> what one takes the requirements for genuine forms of perception to be, and this, in turn, hangs together with what one takes as one's model or paradigm of genuine perception. Wittgenstein seems to have been right when he said that, in the description of aspect perception, 'problems about the concept of seeing come to a head' (Wittgenstein 1982: §172).

Although I have not argued for or against moral perception, I have argued that critics of moral perception may have been too quick in saying that we cannot see moral value. According to Wittgenstein, concepts often have many more uses than philosophers have imagined (Wittgenstein 1969: 50–51, 2005b: 307, 2009a: §38, §47, §108), and the concept of seeing is one of these concepts (Wittgenstein 2009b: §111). Critics of moral perception have typically taken one or some of these uses (seeing a certain shape, seeing a tree) as obviously paradigmatic. But why would these be the only right models for thinking about perception? Why would they be more fundamental than seeing emotion? The perception of moral value and the perception of emotion are remarkably similar. Critics of moral perception will either have to point out crucial differences or to deny that we can see emotion. That will be no easy task, and I am inclined to agree with Wittgenstein:

> For I also say that I see the look that you cast at someone else. And if someone wanted to correct me and say that I don't really *see* it, I would take that for pure stupidity. [...] For 'naïve language', that is to say our naïve, normal way of expressing ourselves, does not contain any theory of seeing – does not show you a *theory* but only a *concept* of seeing. (Wittgenstein 1981: §223)

Chapter 7 builds in several ways upon 'Moral Modesty, Moral Judgment and Moral Advice. A Wittgensteinian Approach' (De Mesel 2014). In that article, I took some of Wittgenstein's ethical views (Wittgenstein 2003: 85; Bouwsma 1986: 45; Rhees 1965: 22; Waismann 1965: 16) as a starting point for discussion about whether we should express moral judgment when asked for moral advice. In Chap. 8, I have a closer look at questions for moral advice, inspired by Wittgenstein's later philosophy rather than by his explicitly stated ethical views. I compare moral and non-moral versions of the question 'Should I do x or y?' and conclude that, while non-moral questions of that form typically ask for answers of the form 'You should do x/y' or 'Do x/y', moral questions often do not ask for such answers. Again, the conclusion is that all parties in a contemporary debate share an unwarranted assumption: debates about moral advice usually center on the question of *what* moral advice of the form 'You should do x/y' or 'Do x/y' should be given in response to the question 'Should I do x/y?', rather than on the question of *whether* any advice of that form is asked for.

According to my Wittgensteinian diagnosis, moral philosophers have tended to overlook the differences between moral and non-moral versions of 'Should I do x or y?' because they have been misled by analogies between the forms of expression in different regions of our language. Questions of the form 'Should I do x or y?' may be questions for moral advice or questions of a non-moral kind. Because in both cases the questions look the same, philosophers have treated them in similar ways, mainly by understanding questions for moral advice of the form 'Should I do x or y?' on the model of non-moral questions. However, these questions function differently, they play different roles in our lives and are embedded in our practices in different ways: the contexts in which they are asked and the kind of answers that they ask for are not as similar as philosophers have suggested.

The comparison between moral and non-moral variants of the question 'Should I do x or y?' has allowed me to distinguish some family traits of moral questions of that form. The list is not meant to be exhaustive. Rather, it is meant to show how, through a detailed comparison between two cases that look rather similar on the surface (two variants of 'Should I do x or y?'), some characteristics of questions for moral advice come to the fore. These characteristics are that moral questions of the form 'Should I do x or y?' (1) are typically asked in a situation in which both parties in the conversation know each other personally and have a shared history, (2) ask for serious consideration, engagement, commitment and interest, (3) ask for an answer recognizing their delicacy, (4) ask for an answer offering a deeper understanding of the meaning of the alternatives and (5) asks for an answer recognizing that a moral decision is non-accidentally and inescapably the agent's to make.

Chapter 8 is in many ways related to Chap. 7. Its main question is: what is a moral argument? An answer that is often suggested is that any argument dealing with moral issues, such as the permissibility of killing in certain circumstances, is a moral argument. I call this the thin sense of 'moral argument'. Arguments that we find in normative and applied ethics are all moral in this sense. They often fail, though, to be moral in other respects. I discuss four ways in which morality can be absent from moral arguments in the thin sense. (1) The argument involves or leads to the denial of a moral certainty. (2) The argument displays no participant attitude. (3) The argument offers no deepened understanding of the problem. (4) The argument denies of ignores the personal character (in three different senses) of moral problems. If thin moral arguments suffer from an absence of morality in at least one of these ways, they are not moral arguments in, what I will call, the thick sense of 'moral argument'. A moral argument in the thick sense is an argument that could possibly qualify as a proper response to a moral problem, an argument of which we can imagine that it would help someone struggling with a real moral problem (as a problem in life rather than as a problem in philosophy) to cope with that problem. If an argument in moral philosophy is not thick, then we should not let it weigh upon our considerations of what we morally ought to do, even if we see no independent reason to question the truth of the premises or the logical validity of the argument. The absence of morality in thin arguments means that these arguments will fail to give us a reason to do whatever they claim that we ought to do.

Chapter 8 differs in some respects from the previous chapters in this book. First, it takes inspiration from Wittgenstein's *On Certainty* (Wittgenstein 1974) rather than from the *Philosophical Investigations*. Second, it builds explicitly on the work of philosophers who have been influenced by Wittgenstein, such as Stanley Cavell, Cora Diamond, Raimond Gaita, Sabina Lovibond, Iris Murdoch, Rush Rhees and P.F. Strawson. Although the differences between these authors are not to be neglected, we can speak of a Wittgensteinian tradition in moral philosophy in the sense of 'tradition' explained by Glock, as 'held together *both* by ties of influence *and* by a family of partially overlapping features' (Glock 2008: 223).[10] One of these overlapping features is that all these authors share a primary concern for the way in which traditional moral theories such as utilitarianism and Kantianism tend to misrepresent our moral practices. Authors from the Wittgensteinian tradition are often well-known by other Wittgensteinians, but they are barely present in contemporary mainstream moral philosophy. Nevertheless, I believe that their ideas, views and ways of approaching moral problems could make an interesting contribution there.

What makes the approach of Chap. 8 thoroughly Wittgensteinian are some features that should, by now, sound familiar. First, the chapter shows how philosophers have often been misled by certain analogies between the forms of expression in different regions of our language. The tendency to be misled by analogies between different regions of our language can be avoided or mitigated by emphasizing important disanalogies. In moral matters, we do not have, as in empirical matters, the ideas of fresh evidence and new discoveries. While the characteristic reaction to an advance in scientific knowledge is 'Goodness me, who would have thought of that!', the characteristic response to a moral insight is 'Of course, I should have thought of that!' Second, Chap. 8 points at some widespread, often implicit but unwarranted assumptions in contemporary moral philosophy. One of them is that, as soon as we have a moral issue and an argument about that moral issue, we have a moral argument. Other assumptions are that knowledge of thin moral arguments will help people to deal with real-life moral issues, that a willingness to question everything is a mark of the serious moral person and the good moral philosopher, that ethics is impersonal and impartial, and that ethics is reducible to questions about what we ought to do.

References

Agam-Segal, Reshef, and Edmund Dain, eds. 2018. *Wittgenstein's Moral Thought*. New York: Routledge.

Baker, Gordon. 2004a. *Philosophical Investigations* §122. Neglected Aspects. In *Wittgenstein's Method. Neglected Aspects*, 22–51. Oxford: Blackwell.

———. 2004b. Wittgenstein's Method and Psychoanalysis. In *Wittgenstein's Method. Neglected Aspects*, 205–222. Oxford: Blackwell.

Baker, G.P., and P.M.S. Hacker. 2005. *Wittgenstein: Understanding and Meaning. Volume 1 of An Analytical Commentary on the Philosophical Investigations. Part I: Essays*. 2nd ed., extensively revised by P.M.S. Hacker. Oxford: Blackwell.

Baker, G.P., and P.M.S Hacker. 2009a. *Wittgenstein: Understanding and Meaning. Volume 1 of An Analytical Commentary on the Philosophical Investigations. Part II: Exegesis §§1–184*. 2nd ed., extensively revised by P.M.S. Hacker. Oxford: Wiley-Blackwell.

Baker, G.P., and P.M.S. Hacker. 2009b. *Wittgenstein: Rules, Grammar and Necessity. Volume 2 of An Analytical Commentary on the Philosophical Investigations. Essays and Exegesis of §§185–242*. 2nd ed., extensively revised by P.M.S. Hacker. Oxford: Wiley-Blackwell.

[10] Glock distinguishes a tradition from a school, the latter being 'a tightly knit group based on relatively intimate personal contact and a direct transfer of certain doctrines or methods' (Glock 2008: 220).

Barrett, Cyril. 1991. *Wittgenstein on Ethics and Religious Belief.* Oxford: Blackwell.
Bouwsma, O.K. 1986. *Wittgenstein. Conversations 1949–1951.* Indianapolis: Hackett.
Brandhorst, Mario. 2017. Varieties of Objectivity. Reply to De Mesel. *Philosophical Investigations* 40: 64–81.
Bronzo, Silver. 2012. The Resolute Reading and Its Critics. *Wittgenstein-Studien* 3: 45–80.
Cavell, Stanley. 1999. *The Claim of Reason. Wittgenstein, Skepticism, Morality, and Tragedy.* Oxford: Oxford University Press.
Crary, Alice. 2007a. *Beyond Moral Judgment.* Cambridge, MA: Harvard University Press.
———, ed. 2007b. *Wittgenstein and the Moral Life. Essays in Honor of Cora Diamond.* Cambridge, MA: The MIT Press.
———. 2016. *Inside Ethics. On the Demands of Moral Thought.* Cambridge, MA: Harvard University Press.
Crary, Alice, and Rupert Read, eds. 2000. *The New Wittgenstein.* London: Routledge.
DeAngelis, William. 2007. *Ludwig Wittgenstein. A Cultural Point of View.* Aldershot: Ashgate.
De Mesel, Benjamin. 2014. Moral Modesty, Moral Judgment and Moral Advice. A Wittgensteinian Approach. *International Journal of Philosophy and Theology* 75: 20–37.
———. 2015. Speaking for Oneself. Wittgenstein, Nabokov and Sartre on How (Not) to Be a Philistine. *Philosophy* 90: 555–580.
De Mesel, Benjamin, and James Thompson, eds. 2015. Wittgensteinian Approaches to Moral Philosophy. Special Issue of Ethical Perspectives 22(1).
Diamond, Cora. 1991. *The Realistic Spirit.* Cambridge, MA: The MIT Press.
———. 1996. Wittgenstein, Mathematics and Ethics. Resisting the Attractions of Realism. In *The Cambridge Companion to Wittgenstein*, ed. Hans Sluga and David G. Stern, 226–260. Cambridge: Cambridge University Press.
Edwards, James C. 1982. *Ethics Without Philosophy. Wittgenstein and the Moral Life.* Tampa: University Presses of Florida.
Gaita, Raimond. 2002. *A Common Humanity. Thinking About Love and Truth and Justice.* London: Routledge.
———. 2004. *Good and Evil. An Absolute Conception.* 2nd ed. London: Routledge.
Gentner, Dedre. 1983. Structure-Mapping. A Theoretical Framework for Analogy. *Cognitive Science* 7: 155–170.
Gleeson, Andrew, ed. 2002. Ethics in the Light of Wittgenstein. *Special Issue of Philosophical Papers* 31 (3): 293–321.
Glock, Hans-Johann. 2008. *What Is Analytic Philosophy?* Cambridge: Cambridge University Press.
———. 2015. Wittgensteinian Anti-Anti Realism. One 'Anti' Too Many? *Ethical Perspectives* 22: 99–129.
Hacker, P.M.S. 1993a. *Wittgenstein: Meaning and Mind. Volume 3 of An Analytical Commentary on the Philosophical Investigations. Part I: Essays.* Oxford: Blackwell.
———. 1993b. *Wittgenstein: Meaning and Mind. Volume 3 of An Analytical Commentary on the Philosophical Investigations. Part II: Exegesis §§243-427.* Oxford: Blackwell.
———. 1996. *Wittgenstein's Place in Twentieth Century Analytic Philosophy.* Oxford: Blackwell.
———. 2000a. *Wittgenstein: Mind and Will. Volume 4 of An Analytical Commentary on the Philosophical Investigations. Part I: Essays.* Oxford: Blackwell.
———. 2000b. *Wittgenstein: Mind and Will. Volume 4 of An Analytical Commentary on the Philosophical Investigations. Part II: Exegesis §§428–693.* Oxford: Blackwell.
———. 2001. Philosophy. In *Wittgenstein. A Critical Reader*, ed. Hans-Johann Glock, 322–347. Oxford: Blackwell.
Hanfling, Oswald. 2002. Questions and Answers in Wittgenstein's Philosophy. In *Wittgenstein and the Human Form of Life*, 11–22. London: Routledge.
Hermann, Julia. 2015. *On Moral Certainty, Justification and Practice. A Wittgensteinian Perspective.* Basingstoke: Palgrave Macmillan.
Iczkovits, Yaniv. 2012. *Wittgenstein's Ethical Thought.* Basingstoke: Palgrave Macmillan.

Johnston, Paul. 1989. *Wittgenstein and Moral Philosophy*. London: Routledge.
———. 1999. *The Contradictions of Modern Moral Philosophy*. London: Routledge.
Kenny, Anthony. 1982. Wittgenstein on the Nature of Philosophy. In *Wittgenstein and His Times*, ed. Brian McGuinness, 1–26. Oxford: Blackwell.
———. 2004. Philosophy States Only What Everyone Admits. In *Wittgenstein at Work. Method in the Philosophical Investigations*, ed. Erich Ammereller and Eugen Fischer, 173–182. London: Routledge.
Klagge, James C. 2011. *Wittgenstein in Exile*. Cambridge, MA: MIT Press.
Kuusela, Oskari. 2008. *The Struggle against Dogmatism. Wittgenstein and the Concept of Philosophy*. Cambridge, MA: Harvard University Press.
———. 2011. The Development of Wittgenstein's Philosophy. In *The Oxford Handbook of Wittgenstein*, ed. Oskari Kuusela and Marie McGinn, 597–619. Oxford: Oxford University Press.
Kuusela, Oskari, and Benjamin De Mesel. Forthcoming. *Ethics in the Wake of Wittgenstein*. New York: Routledge.
Litwack, Eric B. 2009. *Wittgenstein and Value. The Quest for Meaning*. London: Continuum.
Lovibond, Sabina. 1983. *Realism and Imagination in Ethics*. Oxford: Blackwell.
———. 2002. *Ethical Formation*. Cambridge, MA: Harvard University Press.
Lurie, Yuval. 2012. *Wittgenstein on the Human Spirit*. Amsterdam: Rodopi.
Malcolm, Norman. 1984. *Ludwig Wittgenstein. A Memoir*. 2nd ed. Oxford: Clarendon.
McGinn, Marie. 1997. *Routledge Philosophy Guidebook to Wittgenstein and the Philosophical Investigations*. London: Routledge.
Monk, Ray. 1991. Ludwig Wittgenstein. In *The Duty of Genius*. London: Vintage.
Moyal-Sharrock, Danièle, ed. 2004. *The Third Wittgenstein. The Post-Investigations Works*. Aldershot: Ashgate.
———. 2012. Cora Diamond and the Ethical Imagination. *British Journal of Aesthetics* 52: 223–240.
Mulhall, Stephen. 2002. Ethics in the Light of Wittgenstein. *Philosophical Papers* 31: 293–321.
Read, Rupert, and Phil Hutchinson. 2010. Therapy. In *Wittgenstein. Key Concepts*, ed. Kelly Dean Jolley, 149–159. Durham: Acumen.
O'Connor, Peg. 2008. *Morality and Our Complicated Form of Life. Feminist Wittgensteinian Metaethics*. Pennsylvania: Pennsylvania State University Press.
O'Hara, Neil. 2018. *Moral Certainty and the Foundations of Morality*. Cham: Springer.
Rhees, Rush. 1965. Some Developments in Wittgenstein's View of Ethics. *The Philosophical Review* 74: 17–26.
———. 1969. *Without Answers*, ed. D.Z. Phillips. London: Routledge and Kegan Paul.
———. 1999. *Moral Questions*, ed. D.Z. Phillips. Houndmills: Macmillan.
Richter, Duncan. 2002. Whose Ethics? Which Wittgenstein? *Philosophical Papers* 31: 323–342.
Schroeder, Severin. 2006. *Wittgenstein. The Way Out of the Fly-Bottle*. Cambridge: Polity.
Stern, David G. 1991. The Middle Wittgenstein. From Logical Atomism to Practical Holism. *Synthese* 87: 203–226.
Szabados, Béla. 2010. *Ludwig Wittgenstein on Race, Gender, and Cultural Identity. Philosophy as a Personal Endeavor*. Lewiston: Edwin Mellen.
Tilghman, B.R. 1991. *Wittgenstein, Ethics and Aesthetics. The View from Eternity*. Albany: State University of New York Press.
Waismann, Friedrich. 1965. Notes on Talks with Wittgenstein. *The Philosophical Review* 74: 12–16.
Winch, Peter. 1972. *Ethics and Action*. London: Routledge and Kegan Paul.
———. 1987. *Trying to Make Sense*. Oxford: Blackwell.
Wisnewski, Jeremy J. 2007. *Wittgenstein and Ethical Inquiry. A Defense of Ethics as Clarification*. London: Continuum.
Wisnewski, Jeremy J., and Henry Jacoby. 2007. Failures of Sight. An Argument for Moral Perception. *American Philosophical Quarterly* 44: 229–244.

Wittgenstein, Ludwig. 1966. *Lectures and Conversations on Aesthetics, Psychology and Religious Belief*. Oxford: Blackwell.
———. 1969. *The Blue and Brown Books. Preliminary Studies for the 'Philosophical Investigations*. 2nd ed. Oxford: Blackwell.
———. 1974. *On Certainty*. Oxford: Blackwell.
———. 1979a. *Notebooks 1914–1916*. 2nd ed. Oxford: Blackwell.
———. 1979b. *Wittgenstein and the Vienna Circle. Conversations Recorded by Friedrich Waismann*. Oxford: Blackwell.
———. 1980a. *Remarks on the Philosophy of Psychology*. Vol. 1. Oxford: Blackwell.
———. 1980b. *Remarks on the Philosophy of Psychology*. Vol. 2. Oxford: Blackwell.
———. 1980c. *Wittgenstein's Lectures. Cambridge 1930–1932*. Chicago: The University of Chicago Press.
———. 1981. *Zettel*. 2nd ed. Oxford: Blackwell.
———. 1982. *Last Writings on the Philosophy of Psychology. Volume 1. Preliminary Studies for Part II of 'Philosophical Investigations'*. Oxford: Blackwell.
———. 1993. In *Philosophical Occasions 1912–1951*, ed. James C. Klagge and Alfred Nordmann. Indianapolis: Hackett.
———. 1998. *Culture and Value*. Rev 2nd ed. Oxford: Blackwell.
———. 2003. In *Public and Private Occasions*, ed. James C. Klagge and Alfred Nordmann. Lanham: Rowman & Littlefield.
———. 2005a. *Tractatus Logico-Philosophicus*. Trans. C.K. Ogden. London: Routledge.
———. 2005b. *The Big Typescript*. Oxford: Wiley-Blackwell.
———. 2009a. *Philosophical Investigations*. 4th ed., Trans. G.E.M. Anscombe, P.M.S Hacker, and Joachim Schulte. Oxford: Wiley-Blackwell.
———. 2009b. Philosophy of Psychology – A Fragment. In *Philosophical Investigations*, 182–244. 4th ed., Trans. G.E.M. Anscombe, P.M.S Hacker, and Joachim Schulte. Oxford: Wiley-Blackwell.
———. 2014. A Lecture on Ethics. In *Lecture on Ethics*, ed. Edoardo Zamuner, Ermelinda Valentina Di Lascio, and D.K. Levy, 42–51. Oxford: Wiley-Blackwell.
Wright, Crispin. 1988. Moral Values, Projection, and Secondary Qualities. *Proceedings of the Aristotelian Society. Supplementary Volume* 62: 1–26.
Zamuner, Edoardo, Ermelinda Valentina Di Lascio, and D.K. Levy, eds. 2014. *Lecture on Ethics*. Oxford: Wiley-Blackwell.

Part I
Wittgenstein's Later Philosophy

Chapter 2
On Wittgenstein's Comparison of Philosophical Methods to Therapies

Wittgenstein's comparison of philosophical methods to therapies has been interpreted in highly different ways. I identify the illness, the patient, the therapist and the ideal of health in Wittgenstein's philosophical methods and answer four closely related questions concerning them that have often been wrongly answered by commentators. The results of this chapter are, first, some answers to crucial questions: philosophers are not literally ill, patients of philosophical therapies are not always philosophers, not all philosophers qualify as therapists, the therapies are not necessarily to be thought of as psychological therapies and the ideal of health does not consist in the end of philosophy. Second, the chapter shows that the comparison has had a misleading effect, because properties of therapies have been illegitimately projected onto the philosophical methods advanced by Wittgenstein.

2.1 Introduction

In his *Philosophical Investigations*, Wittgenstein compares philosophical methods to therapies: 'There is not a *single* philosophical method, though there are indeed methods, different therapies, as it were' (Wittgenstein 2009a: §133). Many have seen Wittgenstein's comparison between philosophical methods and therapies as the key to a new approach to philosophical problems. Nowadays, Wittgenstein's later philosophy is often characterized as 'therapeutic', and the idea of 'Wittgensteinian therapeutic philosophy' has been developed in a number of different ways. I will try to show that, although many of these ways are surely worth exploring, some of them are based on a misunderstanding or an over-interpretation of Wittgenstein's comparison in the *Investigations*.

De Mesel Benjamin. 2015. On Wittgenstein's Comparison of Philosophical Methods to Therapies. *International Journal of Philosophical Studies* 23: 566–583. Published by Taylor & Francis. See http://www.tandfonline.com/doi/abs/10.1080/09672559.2015.1034749.

B. De Mesel, *The Later Wittgenstein and Moral Philosophy*, Nordic Wittgenstein Studies 4, https://doi.org/10.1007/978-3-319-97619-8_2

Although the explicit comparison of philosophical methods to therapies in §133 is not the only reason why Wittgenstein's late philosophy has come to be regarded as therapeutic, it is important to note that this passage contains the only explicit reference to therapy in Wittgenstein's entire *Nachlass*. Therefore, I take it that every characterization of Wittgenstein's later philosophy which invokes the notion of therapy (if it is not limited to a specific period in the development of Wittgenstein's philosophy prior to or after his work on the *Investigations*) has to be compatible with what is said in the *Philosophical Investigations*. While I do not want to suggest that other works by Wittgenstein (and letters, testimonies, etc.) are not valuable source material, it seems clear to me that, in interpreting §133, greater weight should be afforded to the remarks in the *Philosophical Investigations*, the most authoritative text for the later Wittgenstein's philosophical views.[1] Thus, while I will take quotations from the *Investigations* as evidence for my reading, references to other works are only adduced insofar as I take them to be compatible with Wittgenstein's position in the *Investigations* and are meant to illustrate or clarify that position.

The aim of this chapter is primarily exegetical: I want to point out some traps in interpreting the comparison between philosophical methods and therapies in the *Investigations* and explain why commentators may be (and have been) prone to fall into them. Before starting the discussion, I would like to emphasize two things. First, the authors I discuss often hold similar views, but these views are not exactly the same. In what follows, I will focus on similarities rather than differences, and I may not be able to do full justice to the specific contexts in which each author has developed the views under scrutiny. The authors will be treated as exemplifying certain directions of thought, and I will criticize these directions rather than individual authors. It is not my intention to provide an exhaustive overview of existing interpretations of Wittgenstein's comparison and to evaluate them.

Second, I will not present a full reading of §133 nor provide the single right or complete answer to the question of what Wittgensteinian therapeutic philosophy is. Rather, I will focus on what philosophical therapy in the *Investigations* is certainly *not,* thus leaving open the possibility of reading the comparison (or at least several aspects of it) in different ways. In order to structure and limit the discussion, I will show how answers to four particular questions given by commentators tend to diverge from Wittgenstein's own answers to these questions.

The concept of therapy is intimately related to those of a patient, an illness or a disorder, a therapist and an ideal of health. Therefore, we can gain insight into the comparison between therapies and philosophical methods by asking: if philosophical

---

[1]As a reviewer has remarked, Wittgenstein expressed serious doubts about his work in the Preface of the *Philosophical Investigations*. Why then should we take the *Investigations* as authoritative? The best reason for regarding the *Investigations* as authoritative is the very first sentence of the Preface: 'The thoughts that I publish in what follows are the precipitate of philosophical investigations which have occupied me for the last 16 years.' This makes it quite clear that Wittgenstein saw the *Philosophical Investigations*, for all its defects, as the most authoritative text for his later views. What also makes it authoritative is that Wittgenstein decided to have it published, something he did not decide to do with any other later works.

methods correspond to therapies, what corresponds to the illness or disorder, the patient, the therapist and the ideal of health in philosophy? I will focus on four questions. Regarding the illness, the question will be whether philosophical questions are illnesses (Sect. 2.2). Regarding the patients and therapists, I will investigate whether they are philosophers (Sect. 2.3). Regarding the therapies themselves, I will tackle the question of whether they are psychological (Sect. 2.4). And finally, regarding the ideal of health, I will evaluate the claim that it consists in the end of philosophy (Sect. 2.5). There is no logical necessity behind asking these questions in particular, but the choice is far from arbitrary. I have selected these questions because I believe, first, that trying to answer them will shed light on Wittgenstein's comparison and help to focus discussion about it. Second, they have turned out to be particularly tricky questions to which commentators have given answers that differ considerably from Wittgenstein's. The questions may seem insufficiently refined for now, but I will treat them in more detail in the corresponding sections.

2.2 The Illness: Are Philosophical Questions Illnesses?

Wittgenstein writes: 'The philosopher treats a question; like an illness' (Wittgenstein 2009a: §255). In the comparison of philosophical methods to therapies, philosophical questions or problems stand to philosophical methods as illnesses or disorders stand to therapies. According to Wittgenstein, what is distinctive about philosophical questions or problems is that they arise 'through a misinterpretation of our forms of language' (Wittgenstein 2009a: §111). Words in philosophical questions are typically used out of the (linguistic and non-linguistic) contexts in which they are 'at home' (Wittgenstein 2009a: §116).

An example of a philosophical question in the Wittgensteinian sense is 'Is this object composite?' (Wittgenstein 2009a: §47), when the question is asked without any further explanation. For Wittgenstein, the correct answer is: it depends on what you understand by 'composite', 'and that is not an answer to, but a rejection of, the question'. The question cannot be meaningfully answered when it is asked 'absolutely', without any specification, because a word like 'composite' only makes sense in a language-game in which the difference between simple and composite things is clear. The way to deal with philosophical questions is to show that they make no sense by bringing words 'back from their metaphysical to their everyday use' (Wittgenstein 2009a: §116). That is the aim of philosophical therapies. By, for example, giving us an overview of the use of our words (Wittgenstein 2009a: §122), of how words can and cannot be meaningfully used, the philosophical therapist shows the meaninglessness of philosophical questions. In that way, he ultimately makes these questions disappear. They are not solved by giving answers, but 'dissolved like a lump of sugar in water' (Wittgenstein 2005: 310).[2]

[2]There is some discussion about whether there are significant differences between Wittgenstein's conception of philosophy in *The Big Typescript* and *The Blue and Brown Books* on the one hand

Although I have given nothing more than a rough outline, based on uninterpreted quotations, of what Wittgenstein takes philosophical questions to be, I hope that some similarities between illnesses and philosophical questions have become clear. A philosophical question and an illness both have to be treated in such a way as to make them disappear.[3] But these similarities obviously do not justify the claim that, according to Wittgenstein, philosophical questions *are,* literally, illnesses. First, Wittgenstein, who is known to have been painstakingly precise about the use of words, would probably not have written that 'The philosopher treats a philosophical question, *like* an illness' if he had thought that a philosophical question *is* an illness. Second, an illness is often assumed to be a condition or a state or one's personal experience of that condition or state (see, for example, *Webster's*). It is not clear to me how a question could be a condition or a state or an experience. One could say that an illness or a disease is a disorder (see, for example, the *Oxford English Dictionary*), but how can a question be a disorder?[4]

James Conant writes that, like any analogy, Wittgenstein's comparison 'will be misunderstood if one construes it as involving an identification of the items on the left and right hand side of the analogy (philosophy and illness are the same thing) or a subsumption of the concept of the one item wholly under the concept of the other (philosophy is an illness)' (Conant 2011: 639). Yet the subsumptive interpretation that he describes is exactly the one offered by some commentators (Read and Hutchinson 2010; Fischer 2011a; Baker 2004; see further for details). For example, Rupert Read and Phil Hutchinson write that Wittgenstein sees philosophical problems '*as* mental disturbances – we feel them deeply' (Read and Hutchinson 2010: 153). They suggest that 'sees them as' is equivalent here to 'takes them to be' (Read and Hutchinson 2010: 153), and other passages in their text also support a literal

and *Philosophical Investigations* on the other. James Conant (2011), for example, argues for there being such significant differences, while P.M.S. Hacker (2012) defends the opposite view. I will not take sides in the debate here, but I want to emphasize that, when I quote *The Big Typescript* or *The Blue and Brown Books*, I take these quotes to be illustrative of (and therefore obviously compatible with) Wittgenstein's position in *Philosophical Investigations.* The fact that some quotes from *The Big Typescript* and *The Blue and Brown Books* can be used to illustrate Wittgenstein's conception of philosophy in *Philosophical Investigations* is compatible with both Conant's and Hacker's position.

[3] 'Make them disappear' can mean 'make them stop to exist', but it can also mean 'reveal them to have never been in existence'. Wittgenstein wanted to do the latter with philosophical problems, but philosophical problems are dissimilar to illnesses in this respect, as these are mostly not dealt with by showing that they have never existed. Thus, although there are certainly some similarities between philosophical methods and therapies, much care is needed in spelling them out. This supports my conclusion that there are many aspects of Wittgenstein's methods that are not adequately captured by the therapeutic simile.

[4] Scholars have very different opinions about appropriate definitions of 'illness', 'disease' and 'disorder'. I have decided to stick to the definitions given in dictionaries. These definitions capture what we commonly mean by the terms mentioned, and it would be strange if contemporary discussions about the definition of these terms were needed to understand what Wittgenstein meant by them.

reading.[5] How did they come to think that Wittgenstein takes philosophical problems to be mental disturbances?

There seems to be some textual evidence in *Philosophical Investigations* for the claim that philosophical questions are illnesses, of which I will give two examples. First, the idea that philosophical questions are illnesses often goes hand in hand with the idea, based on what Wittgenstein writes in §133, that philosophical methods are therapies. If philosophical methods are therapies, one easily comes to see the illnesses that they treat as illnesses or disorders typically treated by certain kinds of therapy, for example as mental disturbances. I do, however, not see any reason for claiming that Wittgenstein's philosophical methods are therapies. Wittgenstein writes that philosophical methods are 'as it were' therapies. He compares philosophical methods to therapies, he does not say that philosophical methods *are* literally some kind of therapy.

Second, Wittgenstein writes that philosophical problems are deep disquietudes (Wittgenstein 2009a: §111). There is no 'like' or 'as it were' here. Therefore, one could argue that philosophical problems are some kind of deep disquietude. To support this claim, one could mirror an argument that I provided earlier: 'Wittgenstein, who is known to have been painstakingly precise about the use of words, would probably not have written that "Philosophical problems are deep disquietudes" if he had thought that philosophical problems are *like* deep disquietudes.' This argument, however, fails. While the use of 'like' or 'as it were' excludes a literal, subsumptive interpretation, not using 'like' or 'as it were' leaves room for both a subsumptive as well as a metaphorical interpretation. In order to defend a subsumptive interpretation, one would have to explain how philosophical problems could actually be disquietudes, while disquietudes are commonly understood as feelings. What seems to have prompted Wittgenstein to write that philosophical problems are disquietudes is, roughly, that both keep haunting us, and that we should try to get rid of them. I will not elaborate upon this interpretation because my argument does not hinge upon it. Even if one did defend the subsumptive interpretation that philosophical problems are deep disquietudes, it is not clear why these disquietudes would amount to illnesses. It is perfectly possible to have deep disquietudes and yet not be ill or suffer from mental disturbances or disorders.

None of the examples above justifies the claim that philosophical questions are illnesses, and I know of no examples in PI that would do so. I conclude that, according to Wittgenstein, philosophical questions are not (at least not literally) illnesses.

[5] Take, for example, 'Wittgensteinian philosophy is a quest to find a genuinely effective way of undoing the suffering of minds in torment' (Read and Hutchinson 2010: 150). If a metaphorical reading of Read and Hutchinson's claims is possible at all, I do not see why one would favour a metaphorical rather than a literal reading. At least, the literal reading seems to be the most straightforward one, and if 'mental disturbances' is meant metaphorically, Read and Hutchinson should have done more to make that clear so as to prevent the sentence from being misleading.

2.3 The Patient and the Therapist: Are They Philosophers?

Philosophical therapies are methods for treating philosophical questions, and the latter arise through our misunderstanding of the workings of our language. We might think that we misunderstand language because there is something wrong with language, and that, therefore, language is the patient to be treated.[6] Wittgenstein is radically opposed to this plan (Wittgenstein 2009a: §132). There is nothing wrong with language, it 'is in order as it is' (Wittgenstein 2009a: §98). The problem is that we tend to misuse and misunderstand it, particularly when we are dealing with philosophical questions. Reforming our language will not help, because it could never prevent us from misusing language.

What stands to philosophical methods as patients stand to therapies? The patient is the person having the illness or disorder. As we have seen, the 'illnesses' are philosophical questions. In our comparison, the 'patient' is the person who has philosophical questions. A philosopher is a person whose work is to deal with philosophical questions, and, consequently, a person who often has the (as we have seen: metaphorical) illnesses. Therefore, philosophers definitely qualify as patients. It would be a mistake, however, to think that only professional philosophers deal with philosophical questions. Philosophical questions arise through a misunderstanding of the workings of our language, and may emerge in, for example, psychology and mathematics. These questions are not the privilege of professional philosophers, because language 'has the same traps ready for everyone' (Wittgenstein 2005: 312). The assumptions about private experience that Wittgenstein attacks in the *Investigations* are made 'by professional philosophers and ordinary people alike' (Monk 2005: 91). The fact that the later Wittgenstein considered philosophers as well as non-philosophers to be in the grip of conceptual confusions and, hence, to be in need of philosophical therapies, is interestingly illustrated by his work during the Second World War. In helping doctors to clarify their ideas about 'shock', Wittgenstein helped them to (dis)solve philosophical questions that they were confronted with (Monk 1991: 451). Philosophers are not the only ones who have to deal with conceptual confusions. What distinguishes philosophers is that it is their work to do so and that they are far more often dealing with conceptual confusions than other persons. The difference between philosophers and other people, in this respect, is at most a quantitative difference, and not a qualitative difference between 'the essentially ill philosopher' and 'the essentially healthy ordinary man'. In other words, it is not a difference between the ill (those having the illness) and the healthy (those who do not have it). Wittgenstein's philosophical methods are designed to deal with philosophical questions, and these questions are the inevitable consequences of what Bob Plant has called 'the conceptual vulnerabilities of human

[6] This is Dale Jacquette's view (Jacquette 2014: 266). He truly claims that philosophical problems *exist in* language, but that does not imply (as he seems to think) that the problem *is* language. Moreover, Jacquette's view seems clearly to contradict Wittgenstein 2009a: §98.

beings' (Plant 2004: 235). These vulnerabilities are what Wittgenstein points at when he mentions our 'urge to misunderstand' (Wittgenstein 2009a: §109).

Some commentators understand not the philosophical questions themselves, but the urge to misunderstand the workings of our language as the illness to be cured by philosophical therapies (see Peterman 1992: 43). To counter this claim, I propose to compare it with Wittgenstein 2009a: §254: 'So, for example, what a mathematician is inclined to say about the objectivity and reality of mathematical facts is not a philosophy of mathematics, but something for philosophical *treatment*.' What has to be treated is *what* the mathematician is inclined to say, not his *inclination* to misunderstand.[7] Philosophical questions are dissolved 'through an insight into the workings of our language [...] *despite* an urge to misunderstand them' (Wittgenstein 2009a: §109). Philosophical therapies make philosophical questions disappear, *not* the urge or the temptation to have them. These conceptual vulnerabilities are somehow inextricably linked to our being language-users, and there is no way to treat them by philosophical therapies: 'Human beings are deeply imbedded in philosophical, i.e. grammatical confusions' (Wittgenstein 2005: 311). We are instinctively disposed to misunderstand language (Wittgenstein 1981: §391). Language 'keeps seducing us' and 'humans will continue to bump up against the same mysterious difficulties' (Wittgenstein 2005: 312). The urge to misunderstand the workings of our language is not an illness, just like our inclination to misjudge distances in the dark or our vulnerability to getting a cold are not illnesses. We have them, and we have to deal with them. *Despite* our having them, philosophical questions can be dissolved.

All philosophers are patients from time to time, but not all patients are philosophers. Most commentators, however, focus (consciously or contingently) exclusively on philosophers. This is dangerous, because the idea that only philosophers have philosophical questions, combined with the idea that philosophical questions are illnesses, may lead one to think that philosophers are ill while others are healthy. But, Wittgenstein writes, 'a philosopher is not a man out of his senses, a man who doesn't see what everybody sees' (Wittgenstein 1969: 59). We, language-users, are all susceptible to conceptual confusions, vulnerable to the consequences of misunderstanding certain aspects of our language. The distinction to be made here is not the distinction between philosophers and non-philosophers, but the distinction between philosophical questions and other kinds of questions and how these questions have to be approached.

We often do not distinguish between 'treating an illness' and 'treating a person (having the illness)'. When I have cancer, for example, I can equally well say that 'the cancer is being treated' as that 'I am being treated'. That I am being treated does not imply that the aim of the treatment is to change my personality or that, after a successful treatment, my personality will have changed. Things seem to be different for mental illnesses or disorders. Here, treating the illness or disorder may often

[7] It seems confused, in this respect, to claim that 'to state that what we are tempted to say is "something for philosophical treatment" is to identify such temptations as the objects of clarification' (Kuusela 2008: 44).

be interpreted as treating somebody's personality; the person *as a person* is being treated. If philosophical questions are seen as mental illnesses or disorders (see Sect. 2.2), their treatment could easily be understood as the treatment of somebody's personality. In that sense, philosophical therapies would be aimed at, for example, changing a person's character; indeed, some commentators have interpreted them as such (see, for example, Brand 2000). I have tried to show that philosophical questions, as Wittgenstein understands them, are certainly not mental disorders. Hence, there is no need to understand their treatment as the treatment of somebody's personality or character. (If, as a consequence of philosophical therapies, the patient's personality changes, it will be in very particular respects: her way of seeing, her attitude towards a particular philosophical question or a set of philosophical questions will have changed.) As long as there is no such implication, it makes as much sense to say that philosophical treatment is for people as it makes sense to say that it is for philosophical questions.

Philosophers and ordinary people are only patients *in so far as* and *when* they are having philosophical questions, and having these questions is perfectly normal for any human being. Therefore, comparing a philosopher to a patient suffering from an 'intellectual altitude sickness', as Eugen Fischer does (Fischer 2011a: 49), may be misleading, because an illness or sickness tends to prevent us from leading normal, everyday lives.[8] If one leaves this nuance out, one will easily see a qualitative difference where there is only a quantitative one, which may lead one to think that non-philosophers need no cure. Consequently, the illnesses or disorders to be cured by philosophical therapies will not be thought of as 'connected with [...] the oldest pictures *that are engraved into our language itself*' (Wittgenstein 2005: 311), as Wittgenstein thinks, but as necessarily tied to (a) particular (kind of) language-users. They may then seem to be personal rather than conceptual. This does not mean that it is mistaken to focus on philosophers. After all, Wittgenstein himself did so. Philosophers may well be more responsible for conceptual confusions than non-philosophers, and one does not *have to* mention that non-philosophers are also vulnerable to philosophical problems. But it *is* a problem if one suggests that philosophers are ill while others are not, as Fischer tends to do.

The fact that we are all patients raises the question of who can help us: who is the therapist? Here, Wittgenstein is clear: it is the *philosopher* who treats a question like an illness (Wittgenstein 2009a: §255). It is not because they often have the illness themselves that philosophers are unqualified to treat it. As long as the illness is not a mental disturbance or an illness in the literal sense of the word, philosophers are capable of treating misunderstandings of the workings of our language, including their own. That does not mean that philosophers are *nothing but* therapists, which is what Roy Brand suggests when he claims that 'Wittgenstein, like Socrates, regarded himself as a healer and not as a thinker' (Brand 2000: 1). Even when one interprets

[8] To be sure, Fischer does not deny that philosophers can have a perfectly normal life outside of philosophy. The only thing I want to say here is that the term 'sickness' may lead one to suppose that philosophers do *not* have such a normal life.

'therapist' in a literal way, there is no reason to think, as Brand seems to do, that being a therapist excludes being a thinker.

Do all philosophers qualify as therapists? It should be noted that Wittgenstein uses the word 'philosopher' in at least two different ways. In Wittgenstein 2009a: §116, he writes:

> When philosophers use a word – 'knowledge', 'being', 'object', 'I', 'proposition/sentence', 'name' – and try to grasp the *essence* of the thing, one must always ask oneself: is the word ever actually used in this way in the language in which it is at home? –
>
> What *we* do is to bring words back from their metaphysical to their everyday use.

Wittgenstein contrasts what philosophers do with what 'we' do. There are two categories of philosophers. Only Wittgensteinian philosophers ('we') understand that philosophical questions are based on conceptual confusions. Because they understand the specificity of philosophical questions, Wittgensteinian philosophers see the need for Wittgensteinian methods. They have 'a talent for philosophy', which 'consists in a receptiveness: in the ability to receive a strong and lasting impression from a grammatical fact' (Wittgenstein 2005: 311). These philosophers qualify as therapists.

2.4 The Therapies: Are They Psychological?

Wittgenstein's comparison of philosophical methods to therapies has often been understood as a comparison of philosophical methods to psychotherapies. Examples can be found in Fogelin (1996: 35), Glock (1996: 297) and Read and Hutchinson (2010: 149). James Peterman states that 'the concluding claim of paragraph 133 indicates the analogy to psychoanalysis' (1992: 19).

The identification of therapies with psychoanalytic or psychological therapies is probably prompted by Wittgenstein's well-known sympathy for certain aspects of Freud's psychoanalytic method. Wittgenstein once described himself as a 'disciple of Freud' (Wittgenstein 1966: 41) and admired Freud's 'excellent similes' (Wittgenstein 1993: 107). It is conspicuous, however, that explicit references to Freud, psychoanalysis and psychotherapy are completely absent in *Philosophical Investigations*. Stefan Majetschak remarks that 'explicit references to analogies between psychoanalysis and Wittgenstein's concept of philosophical therapy are only to be found in writings that date from 1930 to 1938' (Majetschak 2010: 162). Those who think it obvious that, in §133, Wittgenstein had psychoanalysis or psychotherapies in mind, leave unexplained why he is not or (in the case of psychoanalysis) no longer explicit about that. According to P.M.S. Hacker, Wittgenstein 'had apparently come to think that the analogy is more harmful than useful, and therefore suppressed it in his writings and teachings' (Hacker 2007: 98–99).[9]

[9] On Wittgenstein's ambivalent relationship to Freud's work and psychoanalysis, see Bouveresse (1995) and Majetschak (2010). For reports about Wittgenstein's anger when it was suggested that his philosophical practice was a kind of psychoanalysis, see Wittgenstein 2012: 390; Bouwsma 1986: 36; Malcolm 1984: 39, 48, 101.

Whether that was indeed the case can be left as an open question here, but it seems clear that the *Philosophical Investigations* do not offer any reasons for the *necessity* of seeing Wittgenstein's therapies as psychotherapies (Majetschak 2010: 168). We should at least allow for the possibility that Wittgenstein deliberately uses 'therapies' instead of 'psychotherapies' in the *Investigations*, and that he had good reasons (see the three examples below) for doing so.

Apart from being unnecessary, understanding Wittgenstein as making a *comparison* between his methods and psychotherapy or psychoanalysis is not problematic as such. After all, Wittgenstein's methods *do* have important things in common with psychotherapy and psychoanalysis (see Hacker 2007: 96–99 and Majetschak 2010: 159–169 for detailed discussion). What is problematic is that, although Wittgenstein's comparison is nothing more than a comparison, it has been interpreted as if his philosophical methods were literally kinds of psychotherapy. As a result, ideas connected with psychoanalysis and psychotherapy which are absent in and incompatible with the *Philosophical Investigations* may infect interpretations of Wittgenstein's methods. I will give three examples.

(1) Thinking that Wittgenstein's philosophical therapies are a kind of psychotherapy reinforces the idea that the illnesses or disorders to be treated by them are illnesses or disorders typically treated by psychotherapy, such as mental disturbances. According to Gordon Baker, 'a philosophical problem *is* an individual's internal conflict' and philosophical problems are 'individuals' *troubled states of mind*' (Baker 2004: 212).[10] As we have seen in Sect. 2.2, they are not.

(2) One could argue that the philosophical questions to be treated by Wittgenstein's methods may not be psychological problems, but that the ways to treat them, the methods themselves, are psychological therapies, because they are aimed at taking away psychological causes of these illnesses. In that sense, the therapist would make our urge to misunderstand the workings of our language (Wittgenstein 2009a: §109) disappear. As we have indicated, though, this urge cannot be cured. It is simply there, and it will always be there, because we are language-users. Wittgenstein explicitly recognizes that, *despite* the urge to misunderstand, philosophical questions, particular misunderstandings, can be treated (Wittgenstein 2009a: §109). This is why work in philosophy 'doesn't have an end' (Wittgenstein 2005: 316): although particular misunderstandings can be treated, new misunderstandings will emerge as long as we are language-users.

(3) One of the dangers of seeing Wittgenstein's therapies as psychotherapeutic methods is that this may lead one to obscure the huge differences that Wittgenstein saw between his methods and methods based on scientific discoveries. Not only did Wittgenstein keep on insisting, throughout his career, that philosophical methods are not scientific and that philosophy is not a science, he

[10] Baker discusses his claims in the particular context of Wittgenstein's 'Dictation for Schlick' (Wittgenstein 2013: 3–83), but adds that 'arguably this particular conception of therapy […] runs through and unifies all of his [Wittgenstein's] later philosophy' (Baker 2004: 219).

also, and more importantly for our purposes here, emphasized that philosophical methods are not *based on science*. Philosophical problems are not dissolved 'by coming up with new discoveries' (Wittgenstein 2009a: §109). Instead, 'the work of the philosopher consists in marshaling recollections for a particular purpose' (Wittgenstein 2009a: §127). Philosophers try to provide 'surveyable representations' (Wittgenstein 2009a: §122) of what we already know. That is why Wittgenstein repeatedly remarks that his investigations are not psychological (Wittgenstein 2009b: §372). In contrast to Wittgenstein's philosophical methods, psychotherapy and psychoanalysis are often taken to be based on science. Freud himself thought that he had established a new science of psychology and regarded psychoanalysis as a scientifically informed practice. To unnecessarily interpret Wittgenstein's therapies as a kind of psychotherapy or psychoanalysis may thus lead one to understand Wittgenstein's methods as in need of a scientific basis.[11]

The mistake mentioned above is most clearly committed by Fischer, insofar as his project can be read as an interpretation or elaboration of Wittgenstein's.[12] Fischer studies therapies for conditions involving delusions and uses insights from cognitive linguistics, cognitive and clinical psychology to shed light on how Wittgensteinian therapeutic philosophy can be practiced (Fischer 2011b). These insights are empirical in nature, but, as we have seen, philosophical questions are typically not dissolved by coming up with new discoveries. Fischer wants to provide therapeutic philosophy with a scientific basis, but as Jacques Bouveresse rightly notes, 'The therapy of philosophical maladies must [...] renounce [...] the consolation of leaning on any scientific foundation' (Bouveresse 1995: 19). New discoveries will generate new confusions, but they will not help to dissolve these confusions or to construct methods for doing so. Philosophical methods are not constructed. Wittgenstein is reported to have said in one of his lectures:

> In science you can compare what you are doing with, say, building a house. You must first lay a firm foundation; once it has been laid it must not again be touched or moved. In philosophy we are not laying foundations but tidying up a room, in the process of which we have to touch everything a dozen times. (Wittgenstein 1980: 24)

Fischer is like Freud, the scientist who is 'building a house'. The scientific underpinnings of Fischer's project are thus deeply problematic from a Wittgensteinian perspective on philosophy.

Did Wittgenstein, then, want to 'keep philosophy pure', as Richard Rorty has argued in 'Keeping Philosophy Pure' (Rorty 1982)? In a sense, he certainly did.

[11] This is not to say that seeing Wittgenstein's therapies as a kind of psychotherapy or psychoanalysis *implies* seeing Wittgenstein's methods as science-based. Psychotherapy or Freudian psychoanalysis do not *have to* be seen as science-based. Many commentators have been able to avoid the danger mentioned here. Yet it remains clear that Fischer has not been able to do so.

[12] Fischer writes that he does not want to engage in any Wittgenstein exegesis, but that he proceeds from a Wittgensteinian inspiration (Fischer 2011a: 23). Even if we assume that Fischer's project is not to be read as an elaboration of Wittgenstein's, it still serves as an example of how Wittgenstein's cannot be elaborated.

Philosophical questions cannot be dissolved by applying scientific or science-based methods and *vice versa*. That is not to say that, in Wittgenstein's view, science and philosophy cannot influence each other. First, although philosophy is not based on science, science influences philosophy: 'Is scientific progress useful to philosophy? Certainly. The realities that are discovered lighten the philosopher's task, imagining possibilities' (Wittgenstein 1982: §807). Second, philosophy influences science. Philosophy deals with conceptual confusions. Whenever it succeeds in providing an overview of the use of our words, it helps us to understand which utterances and questions are meaningful. It articulates 'conceptions and ways of looking at things that scientists may adopt and employ in their work' (Kuusela 2008: 340). Wittgenstein has convincingly shown that the mind, for example, need not be thought of as a thing, a box in which thoughts or pains are located and can be perceived by looking into it (Wittgenstein 2009a: §293). It is evident that such an insight can help scientists. Wittgenstein's philosophical investigations into the foundations of mathematics and the philosophy of psychology are not isolated from mathematics and psychology respectively. On the contrary, they are designed to show precisely how conceptual presuppositions have decisively influenced their development. In that sense, philosophy indeed comes '*before* all new discoveries and inventions' (Wittgenstein 2009a: §126).

The therapies to which Wittgenstein compares his philosophical methods are not necessarily to be thought of as psychological or psychoanalytic therapies. If what his methods have in common with therapies is that they are used to treat problems in order to make them disappear, there is no need to think of psychological therapies alone. Physiological therapies, for example, will do just as well. The same holds if what his methods have in common with therapies is that there are many of them, that there is nothing common to all their instances, but that there is some sort of family-resemblance connection between them (Wittgenstein 2009a: §67). In that sense, the comparison with therapies brings out the later Wittgenstein's methodological pluralism.[13] Yet it is quite natural to think of Wittgenstein's methods as psychological therapies if one assumes that philosophical methods are literally therapies and that philosophical questions are literally illnesses. They do not seem to be candidates for physical illnesses, and so they are taken to be mental ones. Consequently, the therapies employed to treat them are those kinds of therapy used to treat mental illnesses or disorders, among which psychological and psychoanalytic therapies are prominent.

[13] If the comparison to therapies is meant to bring out the later Wittgenstein's methodological pluralism (see also Jacquette 2014: 252), it becomes misleading to ask what the alternative to psychotherapies is supposed to be. We do not have to *specify* 'therapies' in any particular way in order to see what Wittgenstein was driving at when he made the comparison, and doing so may prevent us from seeing Wittgenstein's point because it reduces the plurality in 'therapies'. Wittgenstein may have wanted to emphasize that there are many different philosophical methods, just as there are different kinds of therapies. If we reduce (or think we need to reduce) 'therapies' to a specific kind of therapy, this aspect of the comparison disappears.

2.5 The Ideal of Health: The End of Philosophy?

Wittgenstein compares philosophical methods to therapies. Although most commentators consistently use the singular 'therapy' (Harré 2009 is a notable exception), the plural is of crucial importance. There is not one way to approach philosophical questions, but there is a bewildering variety of ways to do so, 'a series that can be continued in both unforeseen and unforeseeable ways' (Conant 2011: 632). Consequently, the therapies cannot be adequately characterized by describing what they intrinsically consist of or how they all proceed. There will never be a checklist for philosophers, a practical handbook with a list of steps on how to construct philosophical therapies. To try to concretize Wittgenstein's therapies in this way would be to succumb to a 'craving for generality' (Wittgenstein 1969: 17), to 'nourish one's thinking with only one kind of example' (Wittgenstein 2009a: §593).[14]

So how can we clarify what philosophical therapies are? First, by giving examples. In fact, this is precisely what Wittgenstein does in his *Philosophical Investigations* (see the 'mind as a box' example). Second, by making clear what these therapies are not or not necessarily. They are, for example, not scientific or based on science, and they are not necessarily to be thought of as psychological therapies. Third, they may not be united by a common intrinsic essence, but they are united by the kind of questions that they treat (see Sect. 2.2) and the kind of health that they aim to bring about.

What do we want to achieve when we treat philosophical questions? What stands to Wittgenstein's philosophical methods as an ideal of health stands to therapies? Philosophical therapies are aimed at 'the discovery of some piece of plain nonsense' (Wittgenstein 2009a: §119), at making us understand that the philosophical question we have been trying to ask and answer is based on a misunderstanding. When Wittgenstein says that 'philosophical problems should *completely* disappear' (Wittgenstein 2009a: §133), he means that particular philosophical problems can and should be completely dissolved. However, this does not imply that *all* philosophical problems can or will be completely dissolved. When a question or problem has been completely dissolved, there is 'complete clarity' *on that issue*. When a particular question turns out to be *some piece* of plain nonsense, we will no longer ask it or try to answer it. The question no longer haunts us, and at this point we can go on and tackle the next question or break off philosophizing if we want to (Wittgenstein 2009a: §133). Health consists in the question being dissolved, and not in 'an ethical and existential change', as Brand thinks (Brand 2000: 7), although the latter might be the result of the former.

Philosophical questions can be dissolved by applying therapies. Wittgenstein does not suggest, though, that we will ever get to the end of our work, because our urge to misunderstand will make new questions appear and old ones can reappear in

[14] Paul Horwich, for example, tends to succumb to this craving for generality. He lists, under the telling heading 'The General Form of a Philosophical Issue', eight elements or stages which he takes to represent the structure of Wittgensteinian philosophical problems (Horwich 2012: 50).

different guises. New flies will fly into the fly-bottle (Wittgenstein 2009a: §309). We are tidying up a room (see Sect. 2.4), but the room of philosophy is a huge one, and sometimes we think that we are ready, but the room is never tidied up forever, because new things are being brought into it and old dust is constantly falling from the ceiling. We cannot do everything at once, we cannot but proceed on a case-by-case basis, and the fact that the room of philosophy never gets tidied up definitively will often frustrate us. Nevertheless, our work is important: imagine what it would look like if nobody ever tidied up the room!

Questions can be dissolved in philosophy, but the end of philosophy is not within reach. Similarly, we can tidy up parts of a room and it may become unnecessary for some specific individuals to tidy up some specific parts for some time. But that does not mean that specific parts will ever be tidied up definitively or that tidying up the room will become unnecessary. Conant writes that, according to Wittgenstein, 'the work of philosophical elucidation – the true office of philosophy – must remain a forever piecemeal and unfinished task (one which, with respect to its application in detail, must go on indefinitely without ever reaching a final resting place)' (Conant 2011: 639). Our conceptual vulnerabilities, due to our being humans and language-users, make the idea that philosophy could ever come to an end simply inconceivable. Wittgenstein wanted philosophical questions to disappear, but he never declared that he considered the disappearance of *all* philosophical questions to be a genuine possibility. What 'gives philosophy peace' (Wittgenstein 2009a: §133) is not the end of it, but the complete dissolution of particular questions, which results in our (at least temporarily) no longer being troubled by them.

Why, then, has Wittgenstein's ideal of health been associated with the end of philosophy by a number of commentators? John Churchill sees in the later Wittgenstein's conception of philosophy a romantic side, involving the idea that 'philosophy should come to an end with the confusions of its practitioners', and a prosaic side, involving the idea that 'philosophy will not come to an end, for its puzzles have no end' (Churchill 1989: 110). He does not decide between the two options. Rorty attributes to Wittgenstein the idea that philosophy should come to an end and then argues against it: 'Philosophy resembles space and time: it is hard to imagine what an "end" to any of the three would look like' (Rorty 1982: 29). Jacquette claims that Wittgenstein 'wants philosophical questioning […] to disappear altogether', that he proposes a 'way of eliminating philosophy' and 'expects to do away with philosophy' (Jacquette 2014: 258).

What may have prompted these interpretations of the later Wittgenstein is, unsurprisingly, the comparison of philosophical methods to therapies. Although certain therapies (both psychological and physiological) sometimes just aim at making illnesses or disorders more bearable, it seems clear that the concepts of therapy and illness are intimately related to the concept of health. It is common to interpret 'being healthy' as 'no longer being ill, no longer having any illnesses'. Thus, the mistaken idea that philosophical methods are therapies suggests (although it does not imply) the possibility of a state in which we no longer have any 'illnesses', that is, philosophical questions. As we have seen, Wittgenstein does not consider this to be a genuine possibility at all.

2.6 Conclusion

It is true that Wittgenstein's remarks on philosophical methods sometimes amount to what Ray Monk has called 'oracular pronouncements' (Monk 2005: 35). Wittgenstein often offers comparisons without explaining them. This makes them difficult to understand. The most important point of this chapter is that, in order to understand them, the comparisons have to be taken *as comparisons*: they point at certain features that philosophical methods and what they are compared to have in common. The comparison between philosophical methods and therapies is certainly fruitful, but it can be misleading if we forget that Wittgenstein presents 'the model [therapies] as what it is, as an object of comparison – as a sort of yardstick; not as a preconception to which reality *must* correspond' (Wittgenstein 2009a: §131). *That* is, as I hope to have shown, a point that has all too often been overlooked.

The results of this chapter are, at least, some answers to crucial questions. Wittgenstein does not think that philosophers are ill, that philosophical questions are illnesses, that only philosophers are patients of philosophical therapies. Not all philosophers qualify as therapists, Wittgenstein's therapies are not necessarily to be thought of as psychological therapies and their ideal of health does not consist in the end of philosophy. The answers tell us in the first place what philosophical therapies are not. They serve as a warning not to extend the comparison beyond its own limits, not to let everything we associate with 'therapies' or 'illness' influence our interpretation of Wittgenstein without there being evidence in his work for our claims; in short, not 'to make the analogy hold throughout' (Wittgenstein 1969: 7). Analogies can easily mislead us and, to quote Wittgenstein, 'by our method we try to counteract the misleading effect of certain analogies' (Wittgenstein 1969: 28), and 'it is the apparent analogy, and again the lack of analogy, between these cases which causes our trouble' (Wittgenstein 1969: 49). A convincing interpretation of Wittgenstein's comparison of philosophical methods to therapies in the *Philosophical Investigations* has to build on an overall understanding of his methods in the *Investigations* and draws its evidence primarily from the *Investigations*. The comparison should not be interpreted by using the word 'therapies' as a starting point from which properties are being projected onto the philosophical methods advanced by Wittgenstein.

Wittgenstein writes: 'The use of expressions constructed on analogical patterns stresses analogies between cases often far apart. And by doing this these expressions may be extremely useful' (Wittgenstein 1969: 28). The comparison of philosophical methods to therapies is surely useful and important, but it is one comparison among others. It makes sense to talk about Wittgenstein's therapeutic methods, but, at the same time, there are aspects of his methods that are not adequately captured by the therapeutic simile. The comparison does not provide a stable basis on which an understanding of the later Wittgenstein's conception of philosophy can be built. To get a clearer view of the later Wittgenstein's methods, a promising way might be to analyse other comparisons on philosophical methods (e.g. showing a fly the way out of the fly-bottle, arranging books of a library (Wittgenstein 1969: 44), etc.), starting from a small number of interrelated questions.[15] Although a systematic account may never emerge, an overview of Wittgenstein's comparisons and what they do and do not tell us about his methods may prevent us from seeing his conception of philosophy through the glasses of only a single comparison; from, in Wittgenstein's words, being held captive by a picture (Wittgenstein 2009a: §115).

References

Baker, Gordon. 2004. Wittgenstein's Method and Psychoanalysis. In *Wittgenstein's Method. Neglected Aspects*, 205–222. Oxford: Blackwell.

Bouveresse, Jacques. 1995. *Wittgenstein Reads Freud. The Myth of the Unconscious*. Princeton: Princeton University Press.

Bouwsma, O.K. 1986. *Wittgenstein. Conversations 1949–1951*. Indianapolis: Hackett.

Brand, Roy. 2000. Philosophical Therapy. Wittgenstein and Freud. *International Studies in Philosophy* 32: 1–22.

Churchill, John. 1989. Wittgenstein and the End of Philosophy. *Metaphilosophy* 20: 103–113.

Conant, James. 2011. Wittgenstein's Methods. In *The Oxford Handbook of Wittgenstein*, ed. Oskari Kuusela and Marie McGinn, 620–645. Oxford: Oxford University Press.

Fischer, Eugen. 2011a. Diseases of the Understanding and the Need for Philosophical Therapy. *Philosophical Investigations* 34: 22–54.

———. 2011b. *Philosophical Delusion and Its Therapy. Outline of a Philosophical Revolution*. London: Routledge.

Fogelin, Robert. 1996. Wittgenstein's Critique of Philosophy. In *The Cambridge Companion to Wittgenstein*, ed. Hans Sluga and David G. Stern, 34–58. Cambridge: Cambridge University Press.

Glock, Hans-Johann. 1996. *A Wittgenstein Dictionary*. Oxford: Blackwell.

Hacker, P.M.S. 2007. Gordon Baker's Late Interpretation of Wittgenstein. In *Wittgenstein and His Interpreters. Essays in Memory of Gordon Baker*, ed. Guy Kahane, Edward Kanterian, and Oskari Kuusela, 88–122. Oxford: Wiley-Blackwell.

———. 2012. Wittgenstein on Grammar, Theses and Dogmatism. *Philosophical Investigations* 35: 1–17.

Harré, Rom. 2009. Wittgenstein's Therapies. From Rules to Hinges. *New Ideas in Psychology* 27: 118–132.

Horwich, Paul. 2012. *Wittgenstein's Metaphilosophy*. Oxford: Oxford University Press.

Jacquette, Dale. 2014. Later Wittgenstein's Anti-Philosophical Therapy. *Philosophy* 89: 251–272.

Kuusela, Oskari. 2008. *The Struggle Against Dogmatism. Wittgenstein and the Concept of Philosophy*. Cambridge, MA: Harvard University Press.

Majetschak, Stefan. 2010. Psychoanalysis of Grammatical Misinterpretations. The Relationship of Ludwig Wittgenstein with the Work of Sigmund Freud. *Wittgenstein-Studien* 1: 151–170.

Malcolm, Norman. 1984. *Ludwig Wittgenstein. A Memoir*. 2nd ed. Oxford: Clarendon.

Monk, Ray. 1991. *Ludwig Wittgenstein*. In *The Duty of Genius*. London: Vintage.

———. 2005. *How to Read Wittgenstein*. London: Granta Books.

Peterman, James F. 1992. Philosophy as Therapy. In *An Interpretation and Defense of Wittgenstein's Later Philosophical Project*. Albany: State University of New York Press.

[15] That does not mean that Wittgenstein's characterizations of philosophy by means of comparisons should be privileged over characterizations in which no comparisons are used. What I offer here is a suggestion for dealing with Wittgenstein's comparisons. I do not want to downplay the importance of Wittgenstein's remarks on philosophy wherein he does not use any comparisons.

Plant, Bob. 2004. The End(s) of Philosophy. Rhetoric, Therapy and Wittgenstein's Pyrrhonism. *Philosophical Investigations* 27: 222–257.
Read, Rupert, and Phil Hutchinson. 2010. Therapy. In *Wittgenstein. Key Concepts*, ed. Kelly Dean Jolley, 149–159. Durham: Acumen.
Rorty, Richard. 1982. Keeping Philosophy Pure. An Essay on Wittgenstein. In *Consequences of Pragmatism. Essays 1972–1980*, 19–36. Minneapolis: University of Minnesota Press.
Wittgenstein, Ludwig. 1966. *Lectures and Conversations on Aesthetics, Psychology and Religious Belief.* Oxford: Blackwell.
———. 1969. *The Blue and Brown Books. Preliminary Studies for the 'Philosophical Investigations*. 2nd ed. Oxford: Blackwell.
———. 1980. *Wittgenstein's Lectures. Cambridge 1930–1932*. Chicago: The University of Chicago Press.
———. 1981. *Zettel*. 2nd ed. Oxford: Blackwell.
———. 1982. *Last Writings on the Philosophy of Psychology. Volume 1. Preliminary Studies for Part II of 'Philosophical Investigations'*. Oxford: Blackwell.
———. 1993. In *Philosophical Occasions 1912–1951*, ed. James C. Klagge and Alfred Nordmann. Indianapolis: Hackett.
———. 2005. *The Big Typescript*. Oxford: Wiley-Blackwell.
———. 2009a. *Philosophical Investigations*. 4th ed., Trans. G.E.M. Anscombe, P.M.S Hacker, and Joachim Schulte. Oxford: Wiley-Blackwell.
———. 2009b. Philosophy of Psychology – A Fragment. In *Philosophical Investigations*, 182–244. 4th ed., Trans. G.E.M. Anscombe, P.M.S Hacker, and Joachim Schulte. Oxford: Wiley-Blackwell.
———. 2012. In *Wittgenstein in Cambridge. Letters and Documents 1911 – 1951*, ed. Brian McGuinness, 4th ed. Oxford: Wiley-Blackwell.
———. 2013. In *The Voices of Wittgenstein. The Vienna Circle. Ludwig Wittgenstein and Friedrich Waismann*, ed. Gordon Baker. London: Routledge.

Chapter 3
Surveyable Representations, the 'Lecture on Ethics', and Moral Philosophy

I argue that it is possible and useful for moral philosophy to provide surveyable representations (as the later Wittgenstein understands the concept) of moral vocabulary. I proceed in four steps. First, I present two dominant interpretations of the concept 'surveyable representation'. Second, I use these interpretations as a background against which I present my own interpretation. Third, I use my interpretation to support the claim that Wittgenstein's 'A Lecture on Ethics' counts as an example of a surveyable representation. I conclude that, since the lecture qualifies as a surveyable representation, it is *possible* to provide surveyable representations of moral vocabulary. Fourth, I argue that it is *useful* for contemporary moral philosophy to provide surveyable representations, because it may help to dissolve problems in current debates. I provide an example of such a debate, namely, the debate between cognitivists and non-cognitivists.

3.1 Introduction

Wittgenstein did not write much about moral philosophy in his later works, at least not explicitly. Thus, it is tricky to attribute to the later Wittgenstein any clear views on morality. That is not to say, however, that moral philosophy has nothing to learn from Wittgenstein's later philosophy. Indeed, the point of this chapter is to show that moral philosophy has a great deal to learn from Wittgenstein's later writings. However, what it can learn is not in the first place to be found in (a reconstruction of) Wittgenstein's views on moral matters.

According to the later Wittgenstein, philosophy is not about advancing theses. Rather, it is an open-ended set of therapeutic methods for making philosophical

De Mesel Benjamin. 2014. Surveyable Representations, the 'Lecture on Ethics', and Moral Philosophy. *The Nordic Wittgenstein Review* 3: 41–69. See http://www.nordicwittgensteinreview.com/article/view/1922.

B. De Mesel, *The Later Wittgenstein and Moral Philosophy*, Nordic Wittgenstein Studies 4, https://doi.org/10.1007/978-3-319-97619-8_3

problems disappear (Wittgenstein 2009a: §133). Philosophical problems arise 'through a misinterpretation of our forms of language' (Wittgenstein 2009a: §111). That is, philosophical problems are the result of words and sentences being used out of the (linguistic and non-linguistic) practical contexts in which they are 'at home' (Wittgenstein 2009a: §116), which inevitably leads to 'conceptual unclarities' (Wittgenstein 2009b: §202). These unclarities are dissolved 'through an insight into the workings of our language' (Wittgenstein 2009a: §109). We lack that insight because

> [...] we don't have *an overview* of the use of our words. – Our grammar is deficient in surveyability. A surveyable representation produces precisely that kind of understanding which consists in 'seeing connections'. Hence the importance of finding and inventing *intermediate links*.
>
> The concept of a surveyable representation is of fundamental significance for us. It characterizes the way we represent things, how we look at matters. (Is this a 'Weltanschauung'?) (Wittgenstein 2009a: §122)

The concept of a surveyable representation is 'of fundamental significance' because giving a surveyable representation of the use of our words is a philosophical method, employed by the philosopher not merely so as to make himself understood, but also in order to get clear about the matter himself (Wittgenstein 1981: §329). By 'assembling what we have long been familiar with', a surveyable representation does not provide any new information, it just reminds us (Wittgenstein 2009a: §127) of things that we are prone to overlook 'because of their simplicity and familiarity' (Wittgenstein 2009a: §129). Insight into the often overlooked workings of our language will reveal philosophical problems to be the result of conceptual confusions and ultimately make them disappear.

If the significance of Wittgenstein's later philosophy lies in its methods, a central question regarding the later Wittgenstein's significance for *moral* philosophy is this: is it possible and useful to apply Wittgensteinian methods in moral philosophy? A pivotal aspect of that question, and the main question of this chapter, is: is it possible and useful for moral philosophers to give surveyable representations of the *moral* use of our words?[1] Philosophers such as Paul Johnston (1989) and Jeremy Wisnewski (2007) have asked that question before, as part of a broader investigation into the usefulness of Wittgensteinian methods for moral philosophy (and given the significance that Wittgenstein attributes to surveyable representations, in a certain way one *has to* ask it in such an investigation). My focus on surveyable representations is narrower than Johnston's and Wisnewski's, but it also goes deeper. In Sects. 3.2 and 3.3 of this chapter, and in contrast to the scarce attention Johnston and Wisnewski pay to interpretational issues and to the question of what exactly a surveyable inter-

[1] By 'the moral use of our words', I mean first of all the moral use of words that typically occur in moral statements (such as 'good', 'right', 'ought'). In 'a good knife', the use of 'good' is not a moral one. But, as (among others) Cora Diamond has pointed out (Diamond 1996: 251–252), statements can be moral statements without there being any typically 'moral' words in them. Some words do not typically occur in moral statements, but do have a moral use in certain contexts. See, on this point, Chap. 4, Sect. 4.2. I do not want to exclude surveyable representations of the use of these words, although I will focus on the use of typically moral words.

pretation *is*, I will develop an interpretation of the concept of a surveyable representation on the basis of textual evidence and against the background of existing interpretations. Using my interpretation, I will then try to answer my main question. In Sect. 3.4, I will argue that within a Wittgensteinian framework, it is *possible* to give surveyable representations of the moral use of words by providing the example of Wittgenstein's 'Lecture on Ethics'. In Sect. 3.5, I will argue that it is *useful* for contemporary moral philosophy to give such surveyable representations, because it may help to dissolve problems in current debates. Herein, I will also provide an example, namely, the debate between cognitivists and non-cognitivists.

3.2 The Concept of a Surveyable Representation: Two Interpretations

If one thing is for sure about the concept of a surveyable representation, it is, as Gordon Baker has pointed out, that 'though clearly important, Wittgenstein's concept of a perspicuous representation is not itself perspicuous' (Baker 2004: 23).[2] A number of interpretations have been put forward and commentators have acknowledged that their interpretations are not the only possible ones (Baker and Hacker 2005: 332; Hutto 2007: 300). This is not surprising, insofar as Wittgenstein explicitly qualifies only one thing as a surveyable representation in his entire oeuvre: the colour-octahedron (Wittgenstein 1975: §1; see, for a picture of the colour-octahedron, Wittgenstein 1975: §221). The colour-octahedron is a surveyable representation of the grammar of colour because it clarifies how colour-words can and cannot be used meaningfully. It 'wears the rules of grammar on its face' (Wittgenstein 1975: §221): looking at the octahedron, we immediately see that it makes sense to talk about yellowish red, but that it is nonsense to talk about yellowish blue. Although the colour-octahedron is the only unambiguous example that Wittgenstein provides of a surveyable representation, commentators commonly, and rightfully, assume that a surveyable representation does not need to be an image, like the colour-octahedron. This is supported by Wittgenstein's remark that a verbal description can take the place of an image (Wittgenstein 1982: 27, 2009b: §18). But what does such a verbal description look like?

I will introduce two interpretations of the concept of a surveyable representation, Baker and Hacker's being the first. According to their reading, the notion of a surveyable representation can be interpreted in a narrow and a broad sense.

> Narrowly understood, a surveyable representation of the grammar of an expression appears to be a grammatical proposition or a few grammatical propositions that shed enough light

[2]It is not easy to translate the original, German expression *übersichtliche Darstellung*. In the revised fourth edition of the *Philosophical Investigations*, *übersichtlich* has been translated as 'surveyable' 'to preserve the reference to *view* and *surview*' (Wittgenstein 2009a: 252), whereas Anscombe's translation (and consequently, most interpretations) uses 'perspicuous'. On this translation issue, see Baker and Hacker (2005: 308).

> on the matter at hand to dispel illusion and to highlight the grammatical category or role of the expression in question. Broadly understood, a surveyable representation is a synopsis of the grammatical rules for the use of an expression [...] Either way, the selection of the salient rules of grammar is guided by *conceptual* problems that arise in the domain in question. (Baker and Hacker 2005: 332)

Wittgenstein provides numerous examples of surveyable representations in a narrow sense:

> He reminds us that naming is preparatory to the use of a word (Wittgenstein 2009a: §26), but not itself a move in the language-game (Wittgenstein 2009a: §49). He points out that a sample is a paradigm, something with which comparison is made, that it is not *described* by an ostensive definition (Wittgenstein 2009a: §50), hence that it is not a point at which language is 'connected to reality', but belongs to the means of representation; that understanding is not a mental state but is akin to an ability (Wittgenstein 2009a: §150); that to mean something is not the same as to think of it (Wittgenstein 2009a: §187); and so on. (Baker and Hacker 2005: 333)

When it comes to the broad interpretation, however, the only example that Wittgenstein provides is his plan for the treatment of psychological concepts (Wittgenstein 1980a: §895, 1980b: §63, §148, 1981: §472) (Baker and Hacker 2005: 333).

There are two problems with Baker and Hacker's interpretation. First, their distinction between a narrow and a broad interpretation seems unnecessary. In both cases, the selection of salient grammatical rules is guided by conceptual problems, i.e., in both cases a surveyable representation is a set of only those grammatical rules needed to dissolve philosophical problems generated by our misuse of language. One could claim that a broad surveyable representation is more 'complete'. But what does 'completeness' amount to if not to a surveyable representation's capacity to completely dissolve philosophical problems; that is, a capacity that is common to both narrow and broad surveyable representations? There would be a difference if a broad surveyable representation were complete by providing an exhaustive description of everything that could possibly be problematic about the grammar of an expression, i.e., if it would contain an answer to every *possible* unclarity. But this interpretation demands a sense of completeness ('exhaustive', '*everything* that could possibly be problematic') that goes against the spirit of Wittgenstein's later philosophy (see Wittgenstein 2009b: §202, 1981: §440). Baker and Hacker themselves acknowledge this elsewhere, noting that 'the notion of a context-free, purpose-independent conception of the totality of rules for the use of an expression is out of place' (Baker and Hacker 2009: 54; see also Wittgenstein 1982: 21 and Kuusela 2008: 67–68, 80–81, 85, 304). Moreover, such a demand for completeness is not met by Wittgenstein's plan for the treatment of psychological concepts (Wittgenstein 1981: §465) and I am unable to see how it even possibly could be.[3] Thus, I suggest dropping the exhaustive reading of the broad interpretation

[3] One might object that the colour-octahedron dissolves *all possible* problems with the grammar of colour-expressions and is therefore the only example of a surveyable representation in Wittgenstein's oeuvre. If this is true, it is difficult to understand why Wittgenstein did not include in his *Philosophical Investigations* the only example of what he judged to be 'of fundamental

and (because this reading seems to me to be the only way to prevent the broad and narrow interpretations from collapsing into each other) considering the distinction between a narrow and a broad interpretation of 'surveyable representation' unnecessary.

The second problem with Baker and Hacker's interpretation of 'surveyable representation' is that it takes surveyable representations to consist only of grammatical propositions. Hans-Johann Glock agrees with their interpretation insofar that he describes a surveyable representation as 'an enumeration or arrangement of grammatical rules/propositions' (Glock 1996: 280). An example of such a rule, according to Baker and Hacker, can be found in Wittgenstein 2009a: §150:

> The grammar of the word 'know' is evidently closely related to the grammar of the words 'can', 'is able to'. But also closely related to that of the word 'understand'. (To have 'mastered' a technique.)

Compare this with the beginning of Wittgenstein 2009a: §151:

> But there is also *this* use of the word 'know': we say 'Now I know!' – and similarly, 'Now I can do it!' and 'Now I understand!'

Both §150 and §151 produce 'that kind of understanding which consists in "seeing connections"', i.e. connections between the grammar of the words 'know', 'can' and 'understand'. Both §150 and §151 can be said to give us (part of) an overview of the use of our words and help us to dissolve certain philosophical problems, engendered, for example, by the confused ideas that to know something is a mental state and to understand something is a mental process. Therefore, an interpretation in which both §150 and §151 clearly qualify as (parts of) surveyable representations is preferable. But while it is clear that §150 consists of grammatical rules (rules for the use of words), things are quite different in §151. To be sure, I am not saying that Baker, Hacker and Glock would not allow for §151 to be part of a surveyable representation (in case of which, of course, they would have to argue that §151 consists of (or is a) grammatical proposition(s)). What I am saying is that, in some cases, what I think clearly qualifies as (part of) a surveyable representation does not as clearly consist of grammatical propositions. Since there is a great deal of controversy as to what grammatical propositions are (see, for example, Aidun 1981; Schmitz 2006; Schroeder 2009), an account of surveyable representations that is not wholly dependent on a particular conception of grammatical propositions is preferable. One could argue that §151 is a grammatical proposition, but one could also argue as that it gives us an example of the actual use of words and not a rule for the use of words. Why would someone who adopts the latter claim be obliged to refrain from seeing it as (part of) a surveyable representation? Could not an example of the actual use of words help us to achieve an overview of the use of our words?

---

significance'. In any case, an example of a *verbal* surveyable representation seems, in the 'exhaustive' reading of the broad interpretation, impossible to find in Wittgenstein's works. Because Wittgenstein explicitly states that 'a method is now demonstrated by examples' (Wittgenstein 2009a: §133), we can expect him to have given at least some examples of surveyable representations.

The Baker-Hacker interpretation was published in 1980. By 1991, Baker had become convinced that it was seriously misguided. In '*Philosophical Investigations* § 122: Neglected Aspects' (Baker 2004), he develops an interpretation of 'surveyable representation' based on an early version of the *Philosophical Investigations*. According to his interpretation,

> [...] Wittgenstein called 'a perspicuous representation of our grammar' anything which has the function of introducing 'perspicuity' into some aspects of the use of some of 'our words' (i.e. anything which manifestly helps somebody to know his way about by dissolving some philosophical problems which bother him). There is no general restriction on what form a perspicuous representation may take. In particular, it need not be either a diagram (like the colour-octahedron) or an assemblage of grammatical rules for the use of 'our words'. (Baker 2004: 31)

I agree with the later Baker's critique, shared by Oskari Kuusela (2008: 225), that a surveyable representation need not be an assemblage of grammatical rules, as the Baker-Hacker interpretation claims. There are several techniques for introducing perspicuity into the use of our words. A simple language-game, for example, may qualify as a perspicuous representation (Baker 2004: 33). Although he does not refer to Baker, Severin Schroeder (2004) takes up this suggestion by defending the claim that for Wittgenstein, 'synoptic representations' are

> [...] simple, fictitious language-games, like that of buying apples (Wittgenstein 2009a: §1), the builders (Wittgenstein 2009a: §2, §8, §15, §21), the description of coloured squares (Wittgenstein 2009a: §48, §64), ordering someone to fetch composite objects (Wittgenstein 2009a: §60, §62), or the reading of a table according to different schemas (Wittgenstein 2009a: §86). (Schroeder 2004: 148)

Another technique is to compare the use of certain words to the use of others (Baker 2004: 34). By way of similes and analogies, Wittgenstein often tries to break the spell of a certain analogy that 'held us captive' (Wittgenstein 2009a: §115). We think, for example, that sensations are private objects, like beetles in boxes (Wittgenstein 2009a: §293). Bringing these unconscious analogies to light, 'to pass from unobvious nonsense to obvious nonsense' (Wittgenstein 2009a: §464), helps to make our use of words surveyable.

3.3 The Concept of a Surveyable Representation: A Third Interpretation

Two interpretations of the concept of a surveyable representation have been briefly discussed: the Baker-Hacker interpretation and the later Baker's interpretation. I think that the later Baker's amendments to the Baker-Hacker interpretation are justified and I will take them into account in what follows. I will now provide an interpretation of 'surveyable representation' in terms of conditions for something to count as a surveyable representation and techniques to produce surveyable representations. I will develop the interpretation against the background of the two interpretations that I have presented. The systematized (and sometimes deliberately simplified) way in which I will present my interpretation may seem somewhat 'un-Wittgensteinian', but I do not believe that it is, as long as its status is properly understood. It is not 'a preconception to which reality *must* correspond', but a model, 'an object of comparison', 'a sort of yardstick' (Wittgenstein 2009a: §131), and I will use it as such in Sects. 3.4 and 3.5.

My interpretation is based on the idea that Wittgenstein's methods are best understood 'by relating his actual proceeding in the text to his own explanations, and *vice versa*' (Ammereller and Fischer 2004: xvii). This is to say that, if Wittgenstein states that the concept of a surveyable representation is of fundamental significance, there have to be examples of surveyable representations in the *Philosophical Investigations*. Wittgenstein's writings are 'in fact replete with perspicuous representations' (Hutto 2007: 303), because 'Wittgenstein's method, generally speaking, *is* that of the perspicuous presentation of language' (Kuusela 2008: 269, my italics).

In my view, the lack of explicit information on surveyable representations invites a fairly liberal interpretation. Rather than rule out candidates for surveyable representations with the argument that Wittgenstein never explicitly linked them to his concept of surveyable representations (as, for example, Baker and Hacker do with Schroeder's suggestion that fictitious language-games serve as surveyable representations in Baker and Hacker 2005: 329), I am willing to consider everything that is not at odds with the minimal conditions we have as a surveyable representation. What are these minimal conditions? First, a surveyable representation should remedy the problem of our not having an overview of the use of our words. Second, it should produce a kind of understanding which consists in seeing connections, an understanding that should contribute to the dissolution of philosophical problems engendered by our misunderstanding of the workings of our language. These conditions indicate the *problems* that surveyable representations are intended to solve (problem condition) and the *goals* that they are meant to achieve (goal condition).

There is no intrinsic characterization of surveyable representations (Baker 2004: 41; Hutto 2007: 303); that is, there are no intrinsic features that all surveyable representations must have in common. In the *Investigations,* Wittgenstein develops different techniques for producing surveyable representations, four of the most important ones of which I will discuss.[4] These techniques constitute, what I will call, the proceeding condition. For something to count as a surveyable representation, it is not absolutely necessary to satisfy this condition, i.e., to display the use of one or more of these techniques, because I agree with Baker and Hutto that there is no general restriction as to what form a surveyable representation can take. Its satisfying the condition, however, together with its satisfying the two necessary

[4] Most commentators would call them 'methods' (see Baker and Hacker 2005: 290–294), but I would prefer to distinguish 'techniques' from 'methods' here. A philosophical method is a (quite general) way of dealing with philosophical problems, of which 'giving surveyable representations' is an example. Techniques are specific ways to practice a method.

conditions already mentioned (problem condition and goal condition), will speak strongly in favour of its being a surveyable representation.

The first technique, examples of which have been given, is the explicit statement of grammatical rules. Second, Wittgenstein points to our actual use of words (Wittgenstein 1969: 56, 61). I have quoted Wittgenstein 2009a: §151, in which Wittgenstein points at some uses of the word 'know'. Another example is Wittgenstein 2009a: §183:

> We do say 'Now I can go on, I mean I know the formula', as we say 'I can walk, I mean I have the time'; but also 'I can walk, I mean I am already strong enough' [...]

Third, 'our method is not merely to enumerate actual uses of words, but rather deliberately to invent new ones, some of them because of their absurd appearance' (Wittgenstein 1969: 28). The invention of new uses of words and fictional concepts (Wittgenstein 1998: 85) is exemplified in fictitious language-games, examples of which have been mentioned. Wittgenstein constantly asks us to 'suppose', 'think of' or 'imagine' a situation in order to shed light on the ways in which words can and cannot be meaningfully used (Wittgenstein 1969: 61). Fourth, words or sentences are compared to others by way of analogies or similes. Wittgenstein writes that 'What I invent are new *comparisons*' (Wittgenstein 1998: 16) and that

> A philosophical problem can be solved only in the right surrounding. We must give the problem a new surrounding, we must compare it to cases we are not used to compare it with. (Wittgenstein 1993: 457)

Comparisons allow us to see both similarities and differences between something and what it is compared to (between, for example, 'knowing' and 'saying' in Wittgenstein 2009a: §78). They can also help us to understand that certain analogies forced their way into our thinking (for example, the analogy between a sensation and a private object).[5]

Although this list of techniques is by no means exhaustive, these are, in my view, four of the most important (maybe even *the* four most important) techniques used by Wittgenstein in the *Philosophical Investigations* to surveyably represent the use of words.[6] To my knowledge, nobody has yet argued that all of these techniques can be

[5] In Wittgenstein 2009a: §130, Wittgenstein writes that 'language-games stand there as *objects of comparison* which, through similarities and dissimilarities, are meant to throw light on features of our language'. Hence, there is no sharp distinction between presenting fictitious language-games (third technique) and offering comparisons (fourth technique). There is also no sharp distinction between presenting grammatical rules showing how words can and cannot be meaningfully used (first technique) and showing how words are actually being used (second technique). Nevertheless, I think it is useful to draw the distinctions as I have done: we can, for example, show how words are being used without explicitly stating rules (see Wittgenstein 2009a: §151) and not all comparisons are fictitious language-games (take, for example, the analogy between philosophical methods and therapies in Wittgenstein 2009a: §133).

[6] Other techniques include, first, Wittgenstein's asking us to consider how we would teach someone the use of certain expressions and how we have learned them (Wittgenstein 1966: 2, 2009a: §208). A second technique is the invention of different notations, such as Wittgenstein's invention, in the *Tractatus*, of the T/F notation (see Wittgenstein 1982: 98–99, 1969: 23; Baker 2004: 30; Baker and Hacker 2005: 293; Glock 1996: 279). A third technique is giving a picture representing grammatical facts (the colour-octahedron, for example).

used to produce surveyable representations. At least three of them (except for the actual use of words), however, have been explicitly linked with the concept of a surveyable representation by several commentators. First, Glock (1996: 280) and Baker and Hacker (2005: 331) have pointed out that surveyable representations consist of rules for the use of words. Second, Baker (2004: 33) allows for language-games to be surveyable representations. Schroeder (2004: 148) sees language-games as being the only plausible candidates and Hutto (2007: 304) mentions them as 'prime examples' of surveyable representations. Third, the idea of giving a surveyable representation by way of analogies and similes is articulated by Baker (2004: 34) and Kuusela (2008, 233). It is also suggested by G.E. Moore in his notes on Wittgenstein's lectures: 'Moreover, I [Moore] cannot possibly do justice to the extreme richness of illustration and comparison which he [Wittgenstein] used: he was really succeeding in giving what he called a "synoptic" view of things which we all know' (Wittgenstein 1993: 50).

The more it exhibits and combines these techniques and the more it approaches 'the same points [...] afresh from different directions' (Wittgenstein 2009a: 3), the more likely the representation is to be surveyable, i.e. to succeed in dissolving philosophical problems. This is why Wittgenstein often uses the four techniques to surveyably represent the use of one word. An example can be found in Wittgenstein 2009a: §§150–151, where he offers a rule for the word 'know', compares it to 'is able to', provides an example of how it is actually used and asks us to imagine a situation in which the word 'know' can be used. All of these techniques can be said to produce 'that kind of understanding which consists in seeing connections'. In this case, they help to make us see the connection between 'to know' and 'to be able to'. They help us to dissolve philosophical problems engendered by our thinking (mistakenly) that 'knowing something' is a hidden, inner mental state.

My interpretation of surveyable representations can be summarized as follows. Giving surveyable representations is an important method of therapeutic philosophy, a way of getting us to see that the philosophical problem we are struggling with is based on conceptual confusions. A surveyable representation should remedy the problem of our not having an overview of the use of our words and expressions (problem condition). Prime examples of techniques for giving surveyable representations are: showing how words and expressions are actually being used, showing how words and expressions can and cannot be meaningfully used, offering comparisons to highlight differences and similarities between the uses of words and expressions, stating rules for the use of words by way of grammatical propositions (proceeding condition). Surveyable representations produce a kind of understanding which consists in seeing connections. This understanding should contribute to the dissolution of philosophical problems engendered by our misunderstanding of the workings of our language (goal condition). Surveyable representations cannot be complete in an absolute sense. They are complete in a relative sense (relative to a philosophical problem) if they succeed in completely dissolving the philosophical problem we are struggling with. Whether a representation is surveyable, i.e. whether the results of applying different techniques have been represented surveyably, ordered 'in such a manner as to shed light upon the problems' (Baker and Hacker 2009: 257), will ultimately be judged by its succeeding to dissolve philosophical problems.

This interpretation (although it is quite liberal in comparison to existing interpretations) excludes a lot of what Wittgenstein does in the *Tractatus*.[7] The *Tractatus* simply does not display the variety of techniques that Wittgenstein uses in the *Investigations* to approach the same problems from different directions. On the contrary, one could say that in the *Tractatus* Wittgenstein approaches the problem of our not understanding the logic of our language (Wittgenstein 2005: §4.003) from one direction only, i.e., by trying to formulate the general, underlying form of the proposition (Wittgenstein 2005: §6).

3.4 The 'Lecture on Ethics' as a Surveyable Representation

In November 1929, Wittgenstein gave a lecture on ethics to the Heretics Society in Cambridge. At the time, his later philosophy and his new method were only 'just dawning' (Hacker 1996: 85). In his lecture, he 'reiterated the view of the *Tractatus* that any attempt to say anything about the subject-matter of ethics would lead to nonsense' (Monk 1991: 277). It is, therefore, no wonder that commentators have seen the lecture as 'backward-looking' and 'evidently an elaboration of ideas already mooted in the *Tractatus*' (Hacker 1996: 302, 77). Because the concept of a surveyable representation is often understood as central to Wittgenstein's later philosophy alone, it might, at first, be difficult to see how the view that the lecture on ethics qualifies as a surveyable representation can be defended.

Although Wittgenstein's new method was indeed just dawning, the concept of a surveyable representation can already be found in the *Philosophical Remarks*, a collection of notes made by Wittgenstein between February 1929 and April 1930. In fact, the example of the colour-octahedron is taken from these notes. The concept of surveyable representation can be found in another early source, namely, the first part of Wittgenstein's 'Remarks on Frazer's *Golden Bough*', which was written in 1931 (Wittgenstein 1993: 133).[8] In his lectures between 1930 and 1933, Wittgenstein

[7] Not everything, because I do not want to exclude the invention of different notations (see previous footnote) from being a technique for producing surveyable representations.

[8] The 'Remarks on Frazer's *Golden Bough*' concern methods of anthropological hermeneutics, not philosophical methods (Hacker 2001: 75). A surveyable representation, in the sense in which Wittgenstein uses the concept in these remarks, is therefore rightly characterized as 'a particular sort of arrangement of related cases of a phenomenon' (Eldridge 1987: 242). What is surveyably represented is not necessarily the use of our words. Baker has criticized the Baker-Hacker interpretation because, on their reading, 'it is a pleonasm to say that the subject-matter of a perspicuous representation is "the grammar of our language" or "the use of our words"' (Baker 2004: 27). The interpretation of surveyable representations that I have presented tries to clarify the notion as it is used in the *Philosophical Investigations*. Because surveyable representations are explicitly linked in the *Investigations* with giving us an overview of the use of our words and remedying the lack of surveyability of our grammar, I do agree with the statement that is disapprovingly quoted by Baker. Wittgenstein may well have thought that 'hypotheses about historical development or observations about a religious ritual' (Baker 2004: 28) can be identified as surveyable representations, but that is not how he uses the concept in the *Investigations*.

used the term 'synopsis' for *Übersicht* (Baker and Hacker 2005: 308). I have quoted Moore, who said that Wittgenstein 'really succeeded in giving a synoptic view of things'. Moore also notes that 'he [Wittgenstein] said you might say that what is satisfactory in Darwin is not such "hypotheses", but his "putting the facts in a system" – helping us to make a "synopsis" of them' (Wittgenstein 1993: 107). According to Moore,

> He [Wittgenstein] also said that he was not trying to teach us any new facts: that he would only tell us 'trivial' things – 'things which we all know already'; but that the difficult thing was to get a 'synopsis' of these trivialities, and that our 'intellectual discomfort' can only be removed by a synopsis of *many* trivialities – that 'if we leave out any, we still have the feeling that something is wrong'. [...] I imagine that it was in this respect of needing a 'synopsis' of trivialities that he thought that philosophy was similar to Ethics and Aesthetics. (Wittgenstein 1993: 114)

The *Philosophical Remarks,* fragments from the 'Remarks on Frazer's *Golden Bough*' and these quotes from Moore suggest that the concept of a surveyable representation was central to Wittgenstein's thought by the early 1930s. Therefore, it is not unreasonable to suppose that he may have had it in mind at the end of 1929, when he wrote the 'Lecture on Ethics'. Another interesting aspect of Moore's quote is that it links Wittgenstein's synopsis with the methods of ethics and aesthetics. This suggestion has also been articulated by Baker and Hacker. For them, the notion of surveyability is 'implicit in his [Wittgenstein's] fragmentary remarks on ethics and aesthetics' (Baker and Hacker 2005: 307). Both the facts that the notion of surveyable representation may very well have been present in Wittgenstein's thinking at the time he wrote the 'Lecture on Ethics', and that a link between this notion and Wittgenstein's approach to ethics is plausible, are, of course, not arguments for the lecture being a surveyable representation. They do, however, at least allow for the possibility of defending this idea.

In order to decide whether the lecture on ethics qualifies as a surveyable representation, I will use my interpretation of a surveyable representation as an object of comparison. Does the lecture satisfy the aforementioned conditions (the problem condition, the proceeding condition and the goal condition)? The problem of our not having an overview of the use of our words is central to the lecture and, therefore, the problem condition seems to be met. Wittgenstein explicitly states that he wants to 'impress' upon us 'that a certain characteristic misuse of our language runs through *all* ethical and religious expressions' (Wittgenstein 2014: 48). Our misuse of moral words like 'ethics', 'good' or 'right', as well as seemingly non-moral words like 'important', 'miracle', 'safe', 'existence' or 'wondering', generates philosophical problems. What Wittgenstein demonstrates in his lecture is that we are misled by surface similarities in our language. In 'He is a good tennis player', 'good' is used in a relative sense, because the player is good *relative to some predetermined standards* of playing tennis well. The statement 'He is a good tennis player' can be translated in terms of those standards without using the word 'good'. Instead of saying 'He is a good tennis player', we can say 'He usually wins when he plays tennis'. In 'He is a good man', 'good' is used in an absolute sense, because, according to Wittgenstein, this kind of goodness is *not relative to some predeter-*

mined standards. Therefore, there is no possible translation in terms of such standards. The similarity of form between 'He is a good man' and 'He is a good tennis player' misleads us, as it makes us think that what we are saying about the man is akin to what we are saying about the tennis player. But the difference is tremendous: Wittgenstein thought of 'He is a good tennis player' as a meaningful proposition, while he thought that 'He is a good man' was nonsense. This is a clear example of how misunderstandings of the workings of our language generate philosophical problems.

Does the 'Lecture on Ethics' satisfy the proceeding condition? First, we should investigate whether it shows how words are actually being used. Wittgenstein states that '*the first thing that strikes one* about all these expressions [expressions containing words like 'good'] is that each of them is *actually used* in two very different senses' (Wittgenstein 2014: 44, my italics). Not only does Wittgenstein point to how words are actually being used, he also asks us, second, to imagine situations in order to shed light upon the ways in which words can and cannot be meaningfully used.

> *Supposing* that I could play tennis and one of you saw me playing and said 'Well you play pretty badly' and *suppose* I answered 'I know, I am playing badly but I do not want to play any better', all the other man could say would be 'Ah then that's all right'. But *suppose* I had told one of you a preposterous lie and he came up to me and said 'You are behaving like a beast' and then I were to say 'I know I behave badly, but then I do not want to behave any better'. Would he then say 'Ah, then that's all right'? Certainly not; he would say 'Well, you *ought* to want to behave better'. Here you have an absolute judgement of value, whereas the first instance was one of a relative judgement. (Wittgenstein 2014: 44, my italics)

Third, comparisons are everywhere in the lecture on ethics. For instance, Wittgenstein compares 'playing badly' with 'behaving badly': while it makes sense to say that you do not want to play any better, saying that you do not want to behave any better does not (Wittgenstein 2014: 44). He tries to clarify what the word 'ethics' means by using metaphors. He writes that 'I can only describe my feeling by the metaphor, that, if a man could write a book on ethics which really was a book on ethics, this book would, with an explosion, destroy all the other books in the world' (Wittgenstein 2014: 46). He goes on to say that 'Ethics, if it is anything, is supernatural and our words will only express facts; as a teacup will only hold a teacup full of water and [even] if I were to pour out a gallon over it' (Wittgenstein 2014: 46). He also points out analogies and similarities between ethical language, the language of aesthetics (Wittgenstein 2014: 43) as well as religious language (Wittgenstein 2014: 48). He compares a scientific way of looking at things with a way of looking at things as miracles (Wittgenstein 2014: 49–50). Fourth, the lecture contains explicit statements of grammatical rules:

> In fact the word 'good' in the relative sense simply means coming up to a certain predetermined standard. (Wittgenstein 2014: 44)

> [...] every judgement of relative value is a mere statement of facts and can therefore be put in such a form that it loses all the appearance of a judgement of value. (Wittgenstein 2014: 44)

> Now what I wish to contend is, that although all judgements of relative value can be shown to be mere statements of facts, no statement of fact can ever be, or imply, a judgement of absolute value. (Wittgenstein 2014: 45)

> To say 'I wonder at such and such being the case' has only sense if I can imagine it not to be the case. (Wittgenstein 2014: 47)

> 'To be safe' essentially means that it is physically impossible that certain things should happen to me, and therefore it is nonsense to say that I am safe *whatever* happens. (Wittgenstein 2014: 48)

In short, examples of important techniques used by Wittgenstein in the *Philosophical Investigations* abound in the 'Lecture on Ethics'. Different techniques are combined to surveyably represent the use of words. The use of 'good', for example, is surveyably represented by all four techniques. Considering that the text of the lecture is hardly ten pages long, it is no exaggeration to call it a concentration of examples of the later Wittgenstein's techniques. Thus, it obviously meets the proceeding condition.

Does the 'Lecture on Ethics' produce a kind of understanding which consists in seeing connections between the uses of words? It certainly shows how words can be used in a variety of ways, as is illustrated by the examples of 'safe', 'wonder', 'good', 'bad', 'ethics', 'miracle' and the explicit grammatical rules quoted above. It thereby produces an understanding which consists in seeing connections between, for example, ethics, aesthetics and religion, or between certain uses of what are typically taken to be 'moral' words ('good') and seemingly non-moral ones ('wonder'), both of them having a relative and an absolute sense, or between different uses of the same word (relative and absolute sense of 'good').

Does the 'Lecture on Ethics' contribute to the dissolution of philosophical problems engendered by our misunderstanding of the workings of our language? A central claim of the lecture is that ethical statements, i.e., statements in which words are typically used in an absolute sense, are nonsensical statements: the extrapolation of words from relative contexts to absolute contexts results in a violation of grammar. To say that, for example, 'I am safe whatever happens' is to violate a grammatical rule. This runs 'against the boundaries of language', 'against the walls of our cage', and is 'perfectly, absolutely, hopeless' (Wittgenstein 2014: 51). Therefore, ethics cannot be a science and does not contribute to our knowledge in any sense (Wittgenstein 2014: 50–51). The philosophical idea that it does is misguided, and so are, consequently, the philosophical problems generated by that idea. An example of such a problem, mentioned by Wittgenstein, is the problem of finding a description or a definition of 'absolute value'. It is useless to look for such a description, because 'no description that I can think of would do to describe what I mean by "absolute value"', and 'I would reject every significant description that anybody could possibly suggest, *ab initio*, on the ground of its significance' (Wittgenstein 2014: 50). Because the lecture on ethics contributes to the dissolution of philosophical problems generated by a misunderstanding of the workings of our language, the goal condition is met.

Thus, the 'Lecture on Ethics' fulfills all three conditions and thereby qualifies as a surveyable representation of the moral use of words. Why is this important? First, by showing that the lecture is a surveyable representation, we can substantiate the claim that it is possible to work out such representations within a Wittgensteinian framework. If Wittgenstein himself has done so (although he did not present the lecture as such), there are apparently no principled reasons against the application of the notion of a surveyable representation to the moral use of words, although some authors seem to think that there are (see, for instance, Diamond 1996, Kelly 1995 and Richter 1996).[9] The first part of the main question of this chapter, then, can be answered affirmatively. Second, the fact that the lecture is a surveyable representation has remarkable consequences for its interpretation and its place in Wittgenstein's work. One could argue that it is almost redundant to say that Wittgenstein uses Wittgensteinian methods in his own work. As I pointed out at the end of Sect. 3.3, however, giving surveyable representations is a method of the *later* Wittgenstein which excludes a great deal of what Wittgenstein did in the *Tractatus*.[10] The lecture is certainly an elaboration of Tractarian views that are absent in the *Investigations*, for example of the ideas that all moral uses of 'good' have typical features in common (Wittgenstein 2014: 43) and that ethical statements are nonsense (Wittgenstein 2005: §6.421 and Wittgenstein 2014: 50–51). In that sense, the lecture is looking backwards. But that does not guarantee the conclusion, generally endorsed by Wittgenstein interpreters, that it is much closer to the early than to the later Wittgenstein (Hacker and Monk have been mentioned, see also Johnston 1989: 225 and Kelly 1995: 575). Showing that the lecture is a surveyable representation brings out important methodological connections with the later Wittgenstein; that is, Wittgenstein apparently uses a later method in what is traditionally thought to be an earlier work. The lecture is looking backwards when we look at its views on moral matters, but methodologically it is looking ahead. I do not pronounce on the question of whether it is closer to the later than to the early Wittgenstein, but only

[9] I have chosen to show the possibility of surveyable representations of the moral use of words by giving an example (simply because that seems to me the most convincing way), rather than by giving reasons for the possibility and showing the reasons of others against the possibility to be unconvincing. I will not treat in any detail Kelly's, Richter's and Diamond's reasons for rejecting the possibility here. In general, they think that ethics is too different from other philosophical subjects that Wittgenstein deals with (or that ethics is no philosophical subject at all) for it to be treated by the same methods, for instance because ethics is personal or because there is nothing to be said in ethics. See, on this point, Chap. 4.

[10] One could argue that Wittgenstein aims at a surveyable representation of the logic of language in the *Tractatus*. In one sense of 'surveyable representation', this is undoubtedly true, but it is clear that the surveyable representation Wittgenstein aims at in the *Tractatus* is very different from the surveyable representations he provides in the *Investigations*. In this chapter, I present an interpretation of 'surveyable representation' *as used in the* Investigations (see also footnote 8) and argue that *this interpretation* excludes a lot of what Wittgenstein does in the *Tractatus*. The way I present surveyable representations is interwoven with the use of the later Wittgenstein's techniques, and in this sense Wittgenstein does not provide surveyable representations in the *Tractatus* while he does provide them in the 'Lecture on Ethics'. I do not want to deny that Wittgenstein aims at a surveyable representation in another sense in the *Tractatus*.

want to show that there are important connections with the later Wittgenstein and that these are often overlooked if the lecture is labelled as belonging to the 'early' period. In that sense, this section shows that (and how) the notion of surveyable representations developed in Sects. 3.2 and 3.3 can be used to bring out methodological affinities of philosophical texts with the later Wittgenstein's work in general and helps to see, in particular, the 'Lecture on Ethics' as a transitional text in Wittgenstein's oeuvre.

3.5 Are Surveyable Representations Useful for Contemporary Moral Philosophy?

That surveyable representations of the moral use of words are possible and that the 'Lecture on Ethics' qualifies as such a representation is an interesting result for Wittgenstein scholars. It helps to refute claims about the inapplicability of Wittgenstein's later methods to moral philosophy and to reconsider the place of the lecture in Wittgenstein's work. However, if nothing more than a possibility were at stake, moral philosophers would have little reason to care about surveyable representations. Therefore, I will try to show in this section that surveyable representations of the moral use of words, as I have interpreted them in Sects. 3.2 and 3.3, are not only possible but also useful. 'Useful' here means that they may help to dissolve problems in current debates in moral philosophy. To be sure, I do *not* claim, first, that *all* problems in moral philosophy can be dissolved by giving surveyable representations. Surveyable representations help to dissolve conceptual unclarities and not all problems in moral philosophy boil down to such unclarities. More specifically, I think that surveyable representations are more likely to be of help for problems in meta-ethics than for problems in normative or applied ethics (although I do not exclude the latter possibilities).[11] Therefore, I do not advocate the replacement of traditional methods in moral philosophy by the Wittgensteinian method of giving surveyable representations. In some cases, surveyable representations of the moral use of words can serve as a useful methodological addition to traditional methods. Second, I do not claim that the mentioned techniques for producing surveyable representations are new (that moral philosophers have not compared the use of words, have not stated rules for the use of words or have not invented fictitious examples to throw light on how words and expressions can be used). Rather, I think that a conscious combination of these various techniques, a conscious employment of the method of giving surveyable representations which I have tried to make

[11] Someone might object to this chapter by claiming that surveyable representations cannot be of help for problems in moral philosophy, because conceptual investigations will never tell us what to do. Here, however, one (unjustly, I believe) equates 'problems in moral philosophy' with 'moral problems'. See, on that point, Chap. 8, Sect. 8.5.

explicit, is to be recommended.[12] Third, my suggestion does not entail a critique of normative moral philosophy, as if surveyable representations were normatively neutral: one's choice of comparisons, rules and examples of actual or fictitious word-use may well be illustrative of one's normative standpoint.

I will make a case for the usefulness of surveyable representations for contemporary moral philosophy by giving an example of a problem and indicating how surveyable representations could contribute to its dissolution. (I can only provide indications: to dissolve the problem by giving surveyable representations would require a lengthier treatment than can be offered here.) The problem, a popular topic in contemporary meta-ethics, concerns the semantic function of moral judgments. Cognitivists think that moral judgments express beliefs, non-cognitivists think that they express 'non-beliefs' (emotions, prescriptions, etc.). For the problem to be dissolvable by giving surveyable representations, it should first of all fulfill the problem condition; that is, it should be based on a misunderstanding of the workings of language, on our not having an overview of the use of words. Michael Gill (2009) and Walter Sinnott-Armstrong (2009) have convincingly argued that both parties in the debate assume that moral judgments *must* have a semantic function in common: if one moral judgment expresses a belief, then *all* moral judgments must do so. Wittgenstein would have severely criticized the misunderstanding of the workings of language involved here. Why should we assume that there is 'something in common to all the entities which we commonly subsume under a general term' (Wittgenstein 1969 17; see also Wittgenstein 2009a: §§65–67)?[13] Why assume that *all* moral judgments must have a semantic function *in common*, i.e., that they all express beliefs (cognitivism) or that they all express non-belief states (non-cognitivism)? If it can be shown that some moral judgments express beliefs, while others express emotions or prescriptions, the dilemma 'Cognitivism or non-cognitivism?' can be dissolved.

Some authors have recently suggested that the dilemma may be a false one, or that we should at least seriously consider that possibility (Gill 2008; Gill 2009;

---

[12] I do not think that there are many examples of surveyable representations, as I have characterized them in this chapter, to be found in the literature. A good candidate is Georg Henrik von Wright's book *The Varieties of Goodness* (1972). Without explicitly referring to Wittgenstein's concept, von Wright attempts to surveyably represent the grammar of 'good'. He aims at an 'illustration of the multiplicity of uses of the word "good" by means of examples' (von Wright 1972: vii). Hans Oberdiek mentions von Wright's book as an example of Wittgensteinian ethics: 'He [von Wright] shows the fruitfulness of challenging philosophical dogma (such as the distinction between *normative ethics* and *metaethics*) while attending to terms as we use them in thinking about good, duty, and justice. He is especially acute at showing how a *field* of concepts interconnects' (Oberdiek 2009: 194–195). P.M.S. Hacker describes *The Varieties of Goodness* as a 'great work on axiology' in which 'one could discern a Wittgensteinian awareness of "conceptual multiplicity" and a striving for a systematic *surveyable representation* of a large and complex conceptual field, even though the subject-matter was far from Wittgenstein's preoccupations' (Hacker 1996: 144, my italics). I have termed *The Varieties of Goodness* 'a good candidate' for a surveyable representation, because it is unclear to me whether it satisfies the goal condition: it sometimes seems as if bringing out the different uses of 'good' (and not the dissolution of philosophical problems generated by our misusing the term) is von Wright's end.

[13] In the 'Lecture on Ethics', Wittgenstein himself had assumed this: he wanted to point at 'the characteristic features' that moral statements 'all have in common' (Wittgenstein 2014: 43).

Loeb 2008; Sinnott-Armstrong 2009). They have, however, been very careful not to claim that it *is* a false dilemma. For that conclusion to be valid, they think that a great deal more research is necessary. For our purposes, it is important to look at the *methods* that these authors propose for such research. It is conspicuous that they all repeatedly and almost exclusively put their hopes on empirical research, i.e., investigations of actual word-use (Gill 2009: 232; Loeb 2008: *passim*; Sinnott-Armstrong and Wheatley 2012: 371). Their reasoning is as follows: we do not know whether moral judgments have a semantic function in common, and thus we need more examples of actual moral judgments. I do not doubt that these examples may be useful, but more will be necessary. First, one simply cannot gather examples of moral judgments without knowing what 'moral judgment' means, that is, without knowing what it is that one wants to gather examples of. Second, given that one has gathered examples of actual moral judgments in their actual contexts, these examples will have to be ordered, selected and interpreted. The examples alone will not tell us anything about whether the judgments qualify as beliefs, emotions or prescriptions or whether they have a semantic function in common. Therefore, empirical research will simply not be enough.

As an alternative to the empirical method, surveyable representations of the moral use of words are more promising if we want to dissolve the dilemma. Given the proceeding condition, such representations could *include* the empirical evidence mentioned above because that evidence would show how words and expressions are actually being used. But surveyable representations, displaying the use of a variety of techniques, have much more to offer. First, comparing the use of words and expressions may help to understand what moral judgments are and to convert chaotic heaps of data into ordered overviews. Putting certain expressions next to others allows one to see differences and similarities between them and to put together what belongs together.[14] For example: 'We say "He believes that abortion is wrong", but we also say "He just felt that what she did was wrong", and we also speak of moral sentiments. We say that it is true that killing is wrong, but we also say that this truth is different in kind from the truth of empirical beliefs.' Second, it may be helpful to state rules for the use of words. A simple but very important example has already been mentioned: words or expressions do not need to have something in common in order to be commonly subsumed under a general term. By stating this rule, the spell of the picture (Wittgenstein 2009a: §115) that they *must* have something in common can be broken. Third, fictitious examples can make clear how words and expressions can be used.[15] Consider three examples:

(1) John and Jack are walking down the street. John sees a woman being attacked. He wants to help her but he is in a wheelchair. He cannot do anything. He thinks that Jack has not seen the attack and says: 'Somebody's being attacked there.'

---

[14] Wittgenstein compares his method to 'putting together books which belong together', 'taking up some books which seemed to belong together, and putting them on different shelves' (Wittgenstein 1969: 44), and to putting together a jigsaw puzzle (Wittgenstein 1969: 46).

[15] On fictitious examples, thought-experiments and their role in Wittgensteinian moral philosophy, see Diamond (2002).

Jack reacts indifferently: 'She probably asked for it.' Clearly, he does not intend to do anything. John is irritated and pokes him in the back, saying: 'That woman needs help.'

(2) James and Julia are watching television. Julia is pregnant. They are watching a documentary that shows how criminals cut open the bellies of heavily pregnant women and take out their babies to sell them. A pregnant woman is being interviewed. Because the police did nothing to stop the criminals, her husband has set up a neighborhood watch group, but he has recently been killed by the criminals. The woman is mortally afraid as she knows that the criminals have made her and her baby their target. Julia cries: 'That woman needs help!'

(3) Jenny is terminally ill and is suffering unbearable pain. She has requested euthanasia. A committee assesses her request. The advantages and disadvantages of euthanasia and Jenny's specific case are being amply discussed. A member concludes: 'That woman needs help.'

Using these examples, one could argue (as well as on the basis of actual examples or even better, because a set of actual examples may not include three examples of exactly the same sentence) that 'That woman needs help' is a moral judgment in all three cases, that it expresses a prescription in (1), an emotion in (2), and a belief in (3), and that, therefore, our dilemma is a false one.[16]

The suggestions above (although they may be no more than promissory notes) show that a conscious employment of the Wittgensteinian method of providing surveyable representations is more likely to dissolve the 'Cognitivism or non-cognitivism?'-dilemma (goal condition) than the empirical method recommended by Gill, Loeb and Sinnott-Armstrong.[17] That is not to say that the empirical method is bad or useless, but that it is one-sided, that any empirical investigation of moral judgments requires a prior understanding of what moral judgments are and that data have to be selected and ordered. Different techniques for producing surveyable representations allow one to select and order the data in various ways and to approach the same problem from different directions according to Wittgenstein's ideal. Empirical research is not to be excluded, but surveyable representations promise a richer overview of the moral use of words. This richer view, for which I have tried to make a case in this section, makes them particularly useful for contemporary moral philosophy.

In conclusion, I can answer my main question affirmatively: surveyable representations of the moral use of words are both possible within a Wittgensteinian framework and useful for contemporary moral philosophy. This important Wittgensteinian method can and should be transferred to moral philosophy. Wittgenstein himself seems to endorse its transferability:

[16] In footnote 6, I mentioned as an additional technique for producing surveyable representations Wittgenstein's asking how we learn the use of certain words. In Wittgenstein 2009a: §77, he asks how we learnt the use of the word 'good'. Answering this question ('A father saying to his child that it is wrong to do something can best be understood as prescribing the child not to do it.') may contribute to the dissolution of the 'Cognitivism or non-cognitivism?'-dilemma.

[17] For a detailed treatment of these issues, see my forthcoming contribution in Kuusela and De Mesel (2018).

> We are going from one subject-matter of philosophy to another, from one group of words to another group of words. An intelligent way of dividing up a book on philosophy would be into parts of speech, kinds of words. Where in fact you would have to distinguish far more parts of speech than an ordinary grammar does. You would talk for hours and hours on the verbs 'seeing', 'feeling', etc., verbs describing personal experience. We get a peculiar kind of confusion or confusions which comes up with all these words. You would have another chapter on numerals – here there would be another kind of confusion [...] a chapter on 'beautiful', *'good'* – another kind. We get into a new group of confusions; language plays us entirely new tricks. (Wittgenstein 1966: 1, my italics)

> What we want to know, to get a bird's-eye view of [*übersehen wollen*], is the use of the word 'good' [...] (Wittgenstein 1980a: §160).[18]

References

Aidun, Debra. 1981. Wittgenstein on Grammatical Propositions. *The Southern Journal of Philosophy* 19: 141–148.

Ammereller, Erich, and Eugen Fischer, eds. 2004. *Wittgenstein at Work. Method in the Philosophical Investigations*. London: Routledge.

Baker, Gordon. 2004. *Philosophical Investigations* §122. Neglected Aspects. In *Wittgenstein's Method. Neglected Aspects*, 22–51. Oxford: Blackwell.

Baker, G.P., and P.M.S. Hacker. 2005. *Wittgenstein: Understanding and Meaning. Volume 1 of An Analytical Commentary on the Philosophical Investigations. Part I: Essays*. 2nd ed., extensively revised by P.M.S. Hacker. Oxford: Blackwell.

———. 2009. *Wittgenstein: Rules, Grammar and Necessity. Volume 2 of An Analytical Commentary on the Philosophical Investigations. Essays and Exegesis of §§185–242*. 2nd ed., extensively revised by P.M.S. Hacker. Oxford: Wiley-Blackwell.

Diamond, Cora. 1996. Wittgenstein, Mathematics and Ethics. Resisting the Attractions of Realism. In *The Cambridge Companion to Wittgenstein*, ed. Hans Sluga and David G. Stern, 226–260. Cambridge: Cambridge University Press.

———. 2002. What if x Isn't the Number of Sheep?' Wittgenstein and Thought-Experiments in Ethics. *Philosophical Papers* 31: 227–250.

Eldridge, Richard. 1987. Hypotheses, Criterial Claims, and Perspicuous Representations. Wittgenstein's 'Remarks on Frazer's *The Golden Bough*'. *Philosophical Investigations* 10: 226–245.

Gill, Michael. 2008. Meta-Ethical Variability, Incoherence, and Error. *Moral Psychology. Volume 2. The Cognitive Science of Morality: Intuition and Diversity* Walter Sinnott-Armstrong, 387–401. Cambridge, MA: MIT Press.

———. 2009. Indeterminacy and Variability in Meta-Ethics. *Philosophical Studies* 145: 215–234.

Glock, Hans-Johann. 1996. *A Wittgenstein Dictionary*. Oxford: Blackwell.

Hacker, P.M.S. 1996. *Wittgenstein's Place in Twentieth Century Analytic Philosophy*. Oxford: Blackwell.

[18] I am grateful to Stefan Rummens for his comments on several versions of this chapter and to Oskari Kuusela and Yrsa Neuman for their comments on a preprint version of the article in the Open Review procedure of *The Nordic Wittgenstein Review*. A version of this chapter was presented at the fourth symposium of the International Wittgenstein Society and the Nordic Wittgenstein Society, 'The Contemporary Significance of Ordinary Language Philosophy' (Åbo Akademi University, Finland, 2013). I am grateful to the audience for their searching questions.

———. 2001. Developmental Hypotheses and Perspicuous Representations. Wittgenstein on Frazer's *Golden Bough*. In *Wittgenstein. Connections and Controversies*, 74–97. Oxford: Clarendon.

Hutto, Daniel. 2007. Getting Clear about Perspicuous Representations. Wittgenstein, Baker and Fodor. In *Perspicuous Presentations. Essays on Wittgenstein's Philosophy of Psychology*, ed. Danièle Moyal-Sharrock, 299–322. Basingstoke: Palgrave Macmillan.

Johnston, Paul. 1989. *Wittgenstein and Moral Philosophy*. London: Routledge.

Kelly, John. 1995. Wittgenstein, the Self, and Ethics. *The Review of Metaphysics* 48: 567–590.

Kuusela, Oskari. 2008. *The Struggle against Dogmatism. Wittgenstein and the Concept of Philosophy*. Cambridge, MA: Harvard University Press.

Kuusela, Oskari, and Benjamin De Mesel, eds. 2018. *Ethics in the Wake of Wittgenstein. Forthcoming*. New York: Routledge.

Loeb, Don. 2008. Moral Incoherentism. How to Pull a Metaphysical Rabbit Out of a Semantic Hat. In *Moral Psychology Volume 2. The Cognitive Science of Morality: Intuition and Diversity*, ed. Walter Sinnott-Armstrong, 356–385. Cambridge, MA: MIT Press.

Monk, Ray. 1991. Ludwig Wittgenstein. In *The Duty of Genius*. London: Vintage.

Oberdiek, Hans. 2009. Wittgenstein's Ethics. Boundaries and Boundary Crossings. In *Wittgenstein and Analytic Philosophy. Essays for P.M.S. Hacker*, ed. Hans-Johann Glock and John Hyman, 175–202. Oxford: Oxford University Press.

Richter, Duncan. 1996. Nothing to Be Said. Wittgenstein and Wittgensteinian Ethics. *The Southern Journal of Philosophy* 34: 243–256.

Schmitz, Barbara. 2006. Grammatical Propositions. *Grazer Philosophische Studien* 71: 227–249.

Schroeder, Severin. 2004. The Demand for Synoptic Representations and the Private Language Discussion - PI 243–315. In *Wittgenstein at Work. Method in the Philosophical Investigations*, ed. Erich Ammereller and Eugen Fischer, 147–169. London: Routledge.

———. 2009. Analytic Truths and Grammatical Propositions. In *Wittgenstein and Analytic Philosophy. Essays for P.M.S. Hacker*, ed. Hans-Johann Glock and John Hyman, 83–108. Oxford: Oxford University Press.

Sinnott-Armstrong, Walter. 2009. Mixed-Up Meta-Ethics. *Philosophical Issues* 19: 235–256.

Sinnott-Armstrong, Walter, and Thalia Wheatley. 2012. The Disunity of Morality and Why It Matters to Philosophy. *The Monist* 95: 355–377.

von Wright, Georg Henrik. 1972. *The Varieties of Goodness*. 4th ed. London: Routledge & Kegan Paul.

Wisnewski, Jeremy J. 2007. *Wittgenstein and Ethical Inquiry*. London: Continuum.

Wittgenstein, Ludwig. 1966. *Lectures and Conversations on Aesthetics, Psychology and Religious Belief*. Oxford: Blackwell.

———. 1969. *The Blue and Brown Books. Preliminary Studies for the 'Philosophical Investigations*. 2nd ed. Oxford: Blackwell.

———. 1975. *Philosophical Remarks*. Oxford: Blackwell.

———. 1980a. *Remarks on the Philosophy of Psychology*. Vol. 1. Oxford: Blackwell.

———. 1980b. *Remarks on the Philosophy of Psychology*. Vol. 2. Oxford: Blackwell.

———. 1981. *Zettel*. 2nd ed. Oxford: Blackwell.

———. 1982. *Wittgenstein's Lectures. Cambridge, 1932–1935*. Amherst: Prometheus Books.

———. 1993. In *Philosophical Occasions 1912–1951*, ed. James C. Klagge and Alfred Nordmann. Indianapolis: Hackett.

———. 1998. *Culture and Value*. Revised 2nd ed. Oxford: Blackwell.

———. 2005. *Tractatus Logico-Philosophicus. Trans. C.K. Ogden. London: Routledge.*

———. 2009a. *Philosophical Investigations*. 4th ed. Trans. G.E.M. Anscombe, P.M.S Hacker, and Joachim Schulte. Oxford: Wiley-Blackwell.

———. 2009b. Philosophy of Psychology – A Fragment. In *Philosophical Investigations, 182–244. 4th ed. Trans. G.E.M. Anscombe, P.M.S Hacker, and Joachim Schulte*. Oxford: Wiley-Blackwell.

———. 2014. A Lecture on Ethics. In *Lecture on Ethics*, ed. Edoardo Zamuner, Ermelinda Valentina, Di Lascio, and D.K. Levy, 42–51. Oxford: Wiley-Blackwell.

Part II
Wittgenstein's Later Philosophy and His Ethical Views

Chapter 4
Wittgenstein, Meta-ethics, and the Subject Matter of Moral Philosophy

Several authors claim that, according to Wittgenstein, ethics has no particular subject matter and that, consequently, there is and can be no such thing as meta-ethics. These authors argue that, for Wittgenstein, a sentence's belonging to ethics is a classification by use rather than by subject matter, and that ethics is a pervasive dimension of life rather than a distinguishable region or strand thereof. In this chapter, I will critically examine the reasons and arguments given for these claims. In my view, a Wittgensteinian perspective does not exclude the possibility of doing meta-ethics and of there being a particular subject matter of moral philosophy. These alleged impossibilities are not the distinguishing marks of Wittgensteinian moral philosophy. What distinguishes the latter from traditional moral philosophy is, rather, its emphasis on alternative ways of thinking about the subject matter of moral philosophy.

4.1 Introduction

In 2002, a special issue of the journal *Philosophical Papers* entitled *Ethics in the Light of Wittgenstein*, was devoted to the relation between Wittgenstein's philosophy and ethics. Several articles in that issue, as well as some others, contain a set of closely interconnected claims that I will critically examine here. First, several authors argue that, according to Wittgenstein, ethics has no particular subject matter (Sect. 4.2). Second, they argue that, because ethics has no particular subject matter, there is and can be no such thing as meta-ethics (Sect. 4.3). Third, according to them, a sentence's belonging to ethics is a classification by use rather than by subject matter (Sect. 4.4). Fourth, they claim that, for Wittgenstein, ethics is a pervasive dimension of life rather than a distinguishable region or strand thereof (Sect. 4.5).

De Mesel Benjamin. 2015. Wittgenstein, Meta-Ethics and the Subject Matter of Moral Philosophy. *Ethical Perspectives* 22: 69–98. Published by Peeters Publishers. See http://poj.peeters-leuven.be/content.php?id=3073458&url=article.

B. De Mesel, *The Later Wittgenstein and Moral Philosophy*, Nordic Wittgenstein Studies 4, https://doi.org/10.1007/978-3-319-97619-8_4

In my view, however, Wittgenstein does not exclude the possibility of doing meta-ethics and of there being a particular subject matter of moral philosophy. What distinguishes Wittgensteinian moral philosophy from traditional moral philosophy is, rather, its emphasis on alternative ways of thinking about the subject matter of moral philosophy (Sect. 4.6).

Before starting the discussion of these claims, I would like to emphasize two things. First, the authors I discuss hold *similar* views, but these views are not *exactly* the same. In what follows, I will focus on similarities rather than differences, and I may not be able to do full justice to the specific contexts in which each author has developed the views under scrutiny. The authors will be treated as exemplifying a certain trend or tendency, and it is the latter that I will criticize. In other words, the authors go their own way, but in the same direction, and I will try to show what is problematic about that direction. Second, the focus will be on views I disagree with. The fact that the balance between disagreement and agreement with several authors tilts, in this chapter, towards disagreement, should not be taken to reflect my overall evaluation of their contributions to Wittgensteinian moral philosophy, as I will make clear in the conclusion.

4.2 'Ethics Has No Particular Subject Matter'

Cora Diamond writes that, according to a Wittgensteinian conception of ethics, 'ethics has no particular subject matter' (Diamond 2000: 153). Similar statements can be found in other work by the same author and in articles by James Conant (2002: 87), Lars Hertzberg (2002: 255), Stephen Mulhall (2002: 293), and Duncan Richter (1996: 252). Why do they think that ethics for Wittgenstein has no particular subject matter?

In 'Wittgenstein, Mathematics, and Ethics' (1996), Diamond criticizes Sabina Lovibond's treatment of ethics in *Realism and Imagination in Ethics* (1983). Lovibond, she says, assumes that what makes ethical sentences ethical is the occurrence of moral predicates. We tend to have the idea of something like a 'moral vocabulary' (Diamond 1996: 252), a limited set of moral words. We tend to think that if a word from this set occurs in a sentence, then that sentence is an ethical sentence, because the word denotes or refers to a moral property and moral properties are the subject matter of ethics. According to Diamond, this idea 'of sentences as about moral subject matter through the presence in them of moral words' (Diamond 1996: 252) is badly mistaken, because those who hold it fail to take into account the Wittgensteinian insight that what a sentence or word means is determined by its use. Hence, the occurrence of a certain word in a sentence does not fix its status as an ethical sentence: 'a sentence's belonging to ethics is a classification by use' (Diamond 1996: 237). Diamond then goes on to show that sentences without any typically moral words ('good', 'right', 'bad', 'wrong', 'duty', 'obligation', 'virtue') can be ethical sentences. She argues convincingly that in certain contexts the insertion of these words may even distort the ethical meaning of a sentence.

First of all, let me say that I agree with Diamond's insistence on the relative irrelevance of typically 'moral' vocabulary. The occurrence of this or that word does not make a sentence into an ethical sentence, nor does the absence of typically moral words or predicates preclude a sentence from being an ethical sentence. The occurrence of ethical words is neither sufficient nor necessary for sentences to be ethical. Whether a sentence or a word is ethical or not is indeed determined by its use. Diamond refers in this context to Simone Weil, in whose work the word 'chance' is an ethical word, and shows how the sentence 'It is only through chance that I was born' can be an ethical sentence (Diamond 1996: 247). The upshot of Diamond's argument is that there is no limited, clearly determinable set of moral words. We cannot make a list and say: these are the moral words, and there are no others. In that respect, Diamond is right to emphasize that moral philosophy has all too often focused on certain words. That there is no *limited* set of moral words does not mean, however, that there is *no* moral vocabulary. Diamond writes:

> There is not, on this view [the *Tractatus* view of ethics], a 'moral vocabulary', a vocabulary through which we mean moral things. If one wanted to give sense to 'moral vocabulary' one might mean: vocabulary we use in saying things that might have application in moral life, but that excludes no words. Since the *Tractatus* might have such use (and was intended to), 'variable', 'Frege' and 'Theory of Types' belong, in this sense, to 'moral vocabulary'. (Diamond 1996: 252–253)

Diamond discusses here the *Tractatus* view of ethics, but she suggests that the later Wittgenstein did not think that there is a moral vocabulary either. In discussing Lovibond's treatment of the later Wittgenstein, Diamond criticizes 'the very idea of "the moral vocabulary"' (Diamond 1996: 251). Why does Diamond suggest that for Wittgenstein there is no moral vocabulary? She may have reasoned as follows: if many seemingly non-moral words may be used in such a way as to acquire a moral sense, a moral vocabulary (if we allow for the idea at all) would be a long list of all kinds of words 'that might have application in moral life', including, for example, 'variable'. If indeed a moral vocabulary is a list of words moral philosophy should focus on and if this list would be very long, then moral philosophy would lose all focus: it would then have to study the use of almost every word, including 'variable', and the very term 'moral philosophy' would become empty. A list of moral words would hardly be distinguishable from a list of all the words in a dictionary.

This argument is problematic. It seems as if Diamond does not sufficiently account here for the Wittgensteinian insight she has been emphasizing: that meaning is use. If what makes 'chance' a moral word in some cases is its specific use in these cases, then a moral vocabulary will include the word 'chance' with a specification of use: in such and such cases, when used so and so, the word 'chance' is a moral word. Why would it be impossible or undesirable to speak of a vocabulary when such specifications are needed? If it would, then no specific vocabulary, be it an ethical one or not, would be possible at all, because there is no reason to assume that Wittgenstein restricted his idea that meaning is use to ethical sentences (on meaning and use, see Sect. 4.4).

Throughout her article, Diamond contrasts ethical sentences with sentences about botany. If we take Diamond's reference to 'meaning is use' seriously, we should say that sentences about botany are about botany in virtue of their use, and there seems to be nothing wrong with that: the occurrence of typical plant-words does not seem to be a necessary condition for a sentence to be about plants, and plant-words may occur in a sentence without it necessarily being about plants. Can we make a limited list of plant-words, saying that these and only these are the words the occurrence of which makes a sentence into a botanical one? We probably cannot. Does this mean that there is no botanical vocabulary? It does not. What it means is that, according to a Wittgensteinian view, any specific vocabulary, be it an ethical or a botanical one, will only count as an ethical or botanical vocabulary if it specifies that certain words are only ethical or botanical when used in such and such ways (which is, after all, not a strange idea: a dictionary contains many such specifications of use). The moral vocabulary will include 'chance' and 'variable' *in their moral uses*. These moral uses can be specified by giving examples: when used thus-and-so, these words are moral words. Hertzberg, echoing Diamond, writes that 'explicit moral locutions are neither necessary nor sufficient for a conversation to be understood as expressive of moral concerns' (Hertzberg 2002: 256). This way of putting things is confused: there are no 'moral locutions' irrespective of their being used as such: if the locutions are moral (which does not mean, according to Diamond and Hertzberg, 'if they contain typically moral words', but 'if they are used so as to be moral'), then they are (almost trivially, I would say) expressive of moral concerns.

The moral vocabulary is not, as Diamond claims, the whole set of words 'that *might* have application in moral life' (my italics), a set that may be thought to exclude no words, but the set of words that *have* such an application, and the application of which is being specified. Hence, the moral vocabulary excludes a great deal, namely all non-moral words (that is, all words in non-moral uses). What Diamond has shown is that it includes more than is commonly thought, that moral word-use is richer, more varied than we tend to think, that the number of moral words may be indefinite because the number of possible moral uses may be indefinite and that moral words may not be sharply distinguishable from non-moral words. Moreover, moral vocabulary may well be more 'open' or 'flexible' than other kinds of vocabulary, consisting of a large number of words that also occur in other vocabularies (such as 'chance'), while botanical vocabulary is more closed ('chance', for example, is not likely to be used in such a way as to acquire a botanical sense). None of these insights, however, leads to the conclusion that there is no moral vocabulary.

Why is the impossibility of a moral vocabulary (according to Diamond's understanding of 'vocabulary' as a limited list of words without any specification as to their use) important to those who claim that ethics has no particular subject matter? It is often thought that, if ethics has a subject matter, this subject matter consists of moral properties (such as, for example, rightness and wrongness). Moral properties are referred to by a moral vocabulary. If there is no such vocabulary, there are no such properties, and hence no subject matter. This line of reasoning is flawed by Wittgensteinian standards. The occurrence of words such as 'right' and 'wrong', that is, the surface grammar of a sentence, may be misleading and does not guarantee that reference is made to moral properties. Similarly, nothing precludes a sentence like 'It is only through chance that I was born', in its moral use, from referring to a moral property such as goodness. Whether there is a moral vocabulary (again, according to Diamond's understanding thereof) or not, and which words it includes, does not necessarily inform us about the existence or non-existence of moral properties.[1]

It may be said that what Diamond wants to emphasize is not so much that, according to Wittgenstein, there is *no* moral vocabulary, but that there is no *limit* to it (Diamond 1996: 248). Mulhall mentions Diamond's 'emphasis upon the absence of pre-given limits to what might count as an ethical use of language' (Mulhall 2002: 310). Similarly, Richter writes that 'there is no limit that can be set to what language could be moral' (Richter 1996: 254). I agree with this, but not with what these authors take to follow from it. They seem to say: *because* there is no limit to what may count as moral language, there is, for Wittgenstein, 'no special ethical arena' (Richter 1996: 254), 'no special area of thought or life or language about which we can do philosophy of a peculiarly moral kind' (Richter 1996: 253). '"Ethics", in Wittgenstein's vocabulary' does not name 'an independent subject matter or separable area of philosophy' (Conant 2002: 87), 'a branch of philosophy with its own proprietary subject matter' (Conant 2002: 90). According to Wittgenstein, morality is not 'an area of its own' (Hertzberg 2002: 268), not a 'separable domain or concern of human existence as such' (Mulhall 2002: 304), not a 'region or strand' of life (Mulhall 2002: 305), and philosophers have been 'plowing a non-existent field' (Hertzberg 2002: 270). I think that two mistakes are involved in concluding this, and both have already been hinted at. First, these authors seem to presuppose that characteristics of moral language (as they understand it) carry over to its subject matter (M1). Because the language is varied, the subject matter is varied. Because almost every sentence can be moral, if it is used in a certain way, moral sentences can be about almost anything. Because there is an indefinite number of ways of using moral language, there is an indefinite number of moral properties. If there is an indefinite number of moral properties and if moral sentences can be about almost anything, almost anything counts as the subject matter of moral philosophy. Hence, the whole idea of a subject matter (like the idea of a moral language or a moral vocabulary) becomes empty. The relation between moral language and its subject matter assumed here is a pictorial relation in which language stands in a one-to-one relation to what it is about, to its subject matter. It does not need to be argued that the later Wittgenstein has shown this model, often associated with the *Tractatus,* to be inaccurate. An indefinite set of moral words can go together with a definite, even a very limited set of moral properties and with a specifiable subject matter.[2]

[1] Diamond seems to acknowledge this objection as a strong objection to her argument, but she does not discuss it, because '[...] there is no space for the kind of discussion that the objection requires' (Diamond 1996: 246).

[2] Diamond explicitly recognizes this (Diamond 1996: 245), but one can explicitly recognize an objection or problem and at the same time not answer it satisfactorily or see its full impact on one's argument.

One could argue that Diamond, Richter, Hertzberg and Mulhall do not presuppose at all that characteristics of moral language carry over to its subject matter. In so doing, one would have to show why they all insist both on the variety of moral language and on there being no subject matter of moral philosophy, if this variety has no obvious consequences for the subject matter or *vice versa*. But even if they do not commit the first mistake (M1), I think they are subject to a second one (M2). Suppose that it can indeed be shown that the moral subject matter, the ethical 'arena', is overwhelmingly varied and far more extensive than most philosophers have thought. Why and how would that amount to there not *being* a moral arena? The mistake here runs parallel to what I said about moral vocabulary: just as the variety of moral word-use in no way indicates that there is no moral vocabulary, the variety of the moral arena in no way indicates that there is no moral subject matter, it only points at the complexity of that subject matter. So while our authors rightfully *problematize* the ideas of a moral vocabulary and a moral subject matter, I do not think that a Wittgensteinian perspective should lead them to *deny* that there is a moral vocabulary and a moral subject matter.

I conclude that, although the authors mentioned above succeed in showing the variety of moral word-use, they fail to indicate why this variety would lead to the absence of a particular subject matter of moral philosophy. Moreover, although they characterize their insights as Wittgensteinian, it can be doubted whether Wittgenstein indeed thought that there is no such subject matter. On the contrary, Wittgenstein seems to have thought just the opposite. In 'A Lecture on Ethics', he writes:

> My subject, as you know, is ethics and I will adopt the explanation of that term which Prof. Moore has given in his book *Principia Ethica.* He says: 'Ethics is the general inquiry into what is good.' Now I am going to use the term 'ethics' in a slightly wider sense […] And to make you see as clearly as possible *what I take to be the subject matter of ethics* I will put before you a number of more or less synonymous expressions each of which could be substituted for the above definition […] if you look through the row of synonyms which I will put before you, you will, I hope, be able to see the characteristic features they all have in common and these are the characteristic features of ethics. (Wittgenstein 2014: 43, my italics)

Wittgenstein explicitly refers to the subject matter of ethics in this passage.[3] This subject matter can be roughly specified: ethics is about what is good (good in an

[3]To say that Wittgenstein thinks that ethics has a subject matter because he writes that he will 'make you see as clearly as possible *what I take to be the subject matter of ethics*' is to take Wittgenstein at his word. It could be objected that this straightforward interpretation of 'A Lecture on Ethics' is not the only possible one. A reviewer remarks that Wittgenstein may have done in the lecture what he did in the *Tractatus*: to dismiss his own statements as sheer nonsense, not containing any intelligible doctrines. For a defence of this interpretation of the lecture, see the introduction in Zamuner, Di Lascio and Levy (2014: 1–41). For a brief discussion of some problems with this interpretation, see De Mesel (2017). I can only say here that I am not convinced by so-called Nonsense Interpretations of the *Tractatus* and that I take, for example, Severin Schroeder's arguments against them (Schroeder 2006: 105–112) to be decisive. It should be clear that, if Wittgenstein did not dismiss his own propositions as sheer nonsense in the *Tractatus,* the proposal that he may have done something similar in the 'Lecture on Ethics' only leads to the view that he does not

absolute sense, not in a relative one; see Chap. 3, Sect. 3.4). The idea of a subject matter can go together with Diamond's observation that one can talk about what is good in indefinitely many ways, but it straightforwardly contradicts the statement that ethics has, according to Wittgenstein, no particular subject matter. At this point, I see only one possible line of defence. The quoted passage occurs in 'A Lecture on Ethics', commonly taken to be written by the early Wittgenstein.[4] The later Wittgenstein changed his mind, and in saying that, according to Wittgenstein, ethics has no particular subject matter, reference was made to the later Wittgenstein. One would not expect Diamond to take this road, however, since she has always emphasized the continuity between the early and the later Wittgenstein. A similar emphasis on the continuity between early and later Wittgenstein can be found in the works of Conant, Mulhall (2002: 304) and Richter, who explicitly states that 'at least in terms of his attitude towards ethics, Wittgenstein's thinking changed little during his career' (Richter 1996: 243). If they are right about the continuity, it seems false to say that, for Wittgenstein, ethics has no particular subject matter.

4.3 'There Is and Can Be No Such Thing as Meta-ethics'

In the opening lines of his article 'Ethics in the Light of Wittgenstein' (2002), Mulhall remarks that

> [...] those of his [Wittgenstein's] followers who have wanted to explore and develop a distinctively Wittgensteinian approach to ethics have tended to do so by applying what they conceive to be his general methodological principles to this particular subject matter. The basic assumption here is that moral philosophy, like philosophy of language and philosophy of mind, is one of the many distinctive branches or sectors into which philosophy as a whole can be divided [...] (Mulhall 2002: 293)

In the rest of his article, Mulhall attacks this 'basic assumption' and defends the claim that, unlike philosophy of language and philosophy of mind, moral philosophy has no particular subject matter. We tend to think that it has, because we think 'that we can demarcate the domain of moral philosophy by reference to a characteristic list or system of distinctively moral concepts' (Mulhall 2002: 304). There is no such list. Therefore, the domain cannot be demarcated. Therefore, there is no particular subject matter.

What exactly is the difference between moral concepts and *distinctively* moral concepts? In order to draw such a distinction, one would have to assume that there are moral concepts that are not distinctively moral. One could think here, I suppose, of goodness, and say that goodness is a moral concept, although not distinctively moral, because we talk, for example, about 'good knives'. Again, what is overlooked

---

dismiss his own propositions as sheer nonsense in the Lecture either. In any case, if it is possible to read the lecture as calling into question its own propositions, then *that* is the interpretation that requires a substantial argument.

[4] For a discussion of this point, see Chap. 3, Sect. 3.4.

here is that meaning is use: in talking about a good knife, one could say that we use the concept of goodness, but goodness is not a moral concept here, because it is not used as such. Hence, every moral concept is a distinctively moral concept. Mulhall, I take it, may also grant this, but emphasize that a list of such concepts will not be *limited.* Therefore, it will not help us to *demarcate* the domain of moral philosophy (Mulhall 2002: 303). The alleged need for demarcation explains Mulhall's use of 'distinctive(ly)'. It also explains why Richter writes that, for Wittgenstein, there is no *special* ethical arena about which we can do philosophy of a *peculiarly* moral kind and why both Diamond and Mulhall do not just say that ethics has no subject matter, but that it has no *particular* subject matter. In the same vein, Conant claims that, according to Wittgenstein, there is no *independent* or *proprietary* subject matter and that moral philosophy is not a *separable* area. The reasoning here seems to run parallel to what made the idea of a moral vocabulary unattractive to these authors: if we cannot demarcate the domain of moral philosophy, then moral philosophy can be about almost anything, and almost anything would count as the subject matter of moral philosophy. One could say that there is a subject matter, but it would include almost anything, which makes it an almost empty thing to say. I agree with the conclusion that, if moral philosophy can be about almost anything, the idea of a subject matter becomes empty. I disagree, however, both with the thought that moral philosophy can be about almost anything (to which I will come back in the following sections) and with the thought that demarcation is needed for there to be a ('particular' or 'special') subject matter.

In order to clarify the latter point, I would like to have a closer look at Mulhall's contrast between moral philosophy, on the one hand, and philosophy of language and philosophy of mind on the other. He refers to 'the incoherence of the idea that moral philosophy is a separable branch or sector of philosophy more generally (on the model of philosophy of language or philosophy of mind)' (Mulhall 2002: 303–304). Mulhall sees a disanalogy: while philosophy of language and philosophy of mind are 'distinctive branches or sectors into which philosophy as a whole can be divided', moral philosophy is not. The use of 'distinctive', 'separable' and 'divided' testifies of a need for demarcation and illustrates the link between subject matter and demarcation in Mulhall's article. 'Demarcation', 'division' and 'distinction' can be understood in two ways here. Perhaps Mulhall is asking for sharp distinctions and boundaries, that is, for *the* one and only demarcation of the one branch from the other, for closed and finite lists of words or concepts, preferably with no overlap, which would clearly distinguish one branch of philosophy from another. I do not see how such a list could be produced for philosophy of language, philosophy of mind or moral philosophy. One could rightly say that philosophy of language is about language and philosophy of mind is about the mind, but there is no obvious disanalogy with Wittgenstein and Moore saying that moral philosophy is the general inquiry into what is good. Just as it is not clear what counts as language and what not, and just as nobody knows what exactly the mind is, moral philosophers are still unclear about what counts as good, about what 'the good' is, about the meaning of 'good'. Moreover, there is considerable overlap between the domains: language can be moral, and moral language belongs both to the subject matter of philosophy of language and to the subject matter of moral philosophy. Because there are no sharp distinctions and boundaries here, we may take Mulhall not to refer to finite or closed lists of concepts or words when he uses 'demarcate'. This would be in line with Wittgenstein pointing out that the absence of sharp boundaries, or the boundaries having blurred edges, in no way indicates that there are no boundaries (Wittgenstein 2009: §71 and §77). Mulhall may not be looking for *the* demarcation, but asks for *a* demarcation. Several demarcations of the same branch are possible and overlap between branches is allowed for. If this is the case, it becomes unclear why Mulhall asks for *distinctively* moral concepts, a *separable* branch of philosophy and a *particular* subject matter. What do these words add here? Can a *particular* subject matter of philosophy of language or mind be *demarcated* by pointing at *distinctive* concepts? Are these *separable* branches of philosophy? This can be doubted, but nevertheless, philosophy of mind and language are commonly taken to have a subject matter (at least by Mulhall himself). These subject matters, language and mind, have been demarcated in various ways and can be demarcated in an indefinite number of ways. The same holds for the subject matter of moral philosophy, a demarcation of which is certainly possible, as the tradition of moral philosophy and Wittgenstein/Moore's 'what is good' show. In short: if a sharp demarcation is needed, none can be given for philosophy of mind, philosophy of language or moral philosophy. If a blurred demarcation is enough, many demarcations can be given for all three subjects. When it comes to demarcation, they all seem to be in very much the same boat.

Even if it is not wholly clear what kind of demarcation Mulhall is asking for, he allows for the possibility of at least some kind of demarcation, but he has two problems with it. First, he writes:

> For how might one demarcate this separate domain of philosophy if not by means of some general characterization of moral reflection and action, conceived of as a conceptually distinct branch or sector of human existence? But any such characterization will, if the above argument is correct, amount not to a description but to a deployment of the concept supposedly under analysis. (Mulhall 2002: 304)

We *can* demarcate the subject matter of moral philosophy after all, but such a demarcation will involve the 'deployment of the concept supposedly under analysis'. Apparently, it is a problem, threatening the usefulness of demarcation, if you demarcate a domain by deploying concepts under analysis in that domain. Here again, however, there is an analogy rather than a disanalogy of moral philosophy with philosophy of language and mind. Will any adequate demarcation of philosophy of language or philosophy of mind not refer to language or mind, that is, to concepts under analysis in these domains? If such demarcations are problematic, then Mulhall's claim that these domains can be demarcated stands in need of explanation: how exactly can this be done, if not by referring to language or mind?

Mulhall's second problem is that 'There is no way of characterizing the subject-matter of moral philosophy that will not itself give expression to one's own ethical interests and concerns' (Mulhall 2002: 303). He may be right, but I can only repeat here that neither the first nor the second of Mulhall's problems amounts to the impossibility of there being a subject matter of moral philosophy. Mulhall himself, however, takes his demarcation problems as support for the conclusion 'that there is and can be no such thing as meta-ethics' (Mulhall 2002: 303; see also Diamond 2000: 162). Because the subject matter of moral philosophy cannot be demarcated, at least not without deploying concepts under analysis in moral philosophy and without giving expression to one's own ethical interests, we have to acknowledge 'the impossibility of moral philosophy as meta-ethics' (Mulhall 2002: 303). This is a striking claim: if meta-ethics were impossible, then what have meta-ethicists been doing? And why do they think that meta-ethics is possible after all?

It is crucial, in this respect, to investigate Mulhall's conception of meta-ethics. First, Mulhall takes the possibility of meta-ethics to depend on the possibility of demarcating the subject matter of moral philosophy. As we have seen, this demarcation problem seems to be no problem at all (at least if, with Mulhall, one refuses to regard it as a problem for philosophy of language and philosophy of mind), regardless of whether one takes the need for demarcation to be a need for sharp distinctions or not. Second, the possibility of meta-ethics depends on the possibility of yet another distinction: Mulhall criticizes what he takes to be an assumption made by Paul Johnston in his *Wittgenstein and Moral Philosophy* (1989), 'that a sharp distinction can be drawn between first-order ethics (advocacy) and meta-ethics (grammatical reminders about the logic of moral concepts)' (Mulhall 2002: 299). According to Mulhall, such a sharp distinction cannot be drawn, because grammatical reminders about the logic of moral concepts will always be expressive of one's ethical interests and concerns. If one demands of meta-ethics that it be normatively neutral, there can be no meta-ethics. If meta-ethics should be impersonal, if it should not involve any personal ethical commitments, there can be no meta-ethics. As Mulhall writes, 'there simply is no space for the straightforward assumption of impersonal linguistic authority' (Mulhall 2002: 319). I am inclined to agree with him, but my question is: who would think that the sharp distinction Mulhall takes to be impossible *can* be drawn? Who would think that meta-ethics *is* normatively neutral and impersonal, not in any way expressive of the meta-ethicist's concerns? Mulhall mentions Charles Stevenson, and refers to 'the meta-ethical enterprise so characteristic of analytic moral philosophy in the first half of the twentieth century (and beyond)' (Mulhall 2002: 303). I do not deny that Stevenson and some others (Alfred Ayer, for example) may have thought the distinction to be sharp. But, just as there are not many moral philosophers nowadays who would say that the subject matter of ethics can be demarcated in a sharp way from the subject matter of other branches of philosophy, I am quite sure that most contemporary meta-ethicists would object to the idea that there is a sharp distinction between normative ethics and meta-ethics. Hence, there is no reason to suppose that meta-ethics, regardless of how one understands it (this may be what 'there can be no meta-ethics' adds to 'there is no meta-ethics'), is impossible without normative neutrality. That point is often explicitly emphasized by contemporary meta-ethicists. In the *Stanford Encyclopedia of Philosophy*, we read:

> In fact, metaethics has seemed to many to offer a crucial neutral background against which competing moral views need to be seen if they are to be assessed properly. Some metaethicists early in the twentieth century went so far as to hold that their own work made no

> substantive moral assumptions at all and had no practical implications. Whether any view that is recognizably still a view about the nature and status of ethics could manage this is dubious. (Sayre-McCord 2012)

In short, Mulhall's reasons for the claims that there is no particular subject matter of moral philosophy and that there can be no meta-ethics do not convince me, and I doubt whether they would convince many meta-ethicists. He puts much emphasis on demarcation, but problems with demarcation pass the question by. If I want to know whether philosophy of language has a subject matter and what that subject matter is, do I have to be able to demarcate 'language'? Can I not do linguistics without knowing where the limits of language lie or biology without knowing where the boundaries of life are? Do I have to know where the boundaries of a country are before I am able to go there and describe, investigate, know and understand what is in it?[5] Mulhall is right, and has Wittgenstein on his side, when he *problematizes* distinctions between moral and non-moral, between meta-ethics and normative ethics, but I do not follow him when he tends to *deny* that there are such distinctions and that these can be characterized. He does not succeed in showing why moral philosophy has no particular subject matter, why Wittgenstein would be wrong in claiming that this subject matter is (roughly) 'what is good' and why there can be no such thing as meta-ethics.

4.4 'A Classification by Use Rather than by Subject Matter'

Our authors all emphasize that what makes a statement moral is not its subject matter but its use. Diamond writes that 'being about good and evil is a matter of use, not subject matter' (Diamond 1996: 244), that the moral character of a sentence 'arises not through its content but from its use on particular occasions' (Diamond 1996: 248) and that 'what makes a statement moral [...] is not its subject matter [...] but its use' (Diamond 1996: 253). Mulhall approvingly mentions Diamond's finding that 'what makes a stretch of discourse [...] moral [...] is a matter of use, not properties denoted or subject-matter' (Mulhall 2002: 306) and Richter does just the same (Richter 1996: 253). If what makes a statement moral is not its subject matter but its use, they suggest, it may be doubted whether ethics has a subject matter at all.

Diamond writes:

> We have seen Wittgenstein's view that a particular sentence might belong to pomology or might belong to mathematics, and *which* it belongs to depends not on what it is apparently about but on its use. Why not consider the question, then, whether a sentence's belonging to ethics is a classification by use rather than by subject matter? (Diamond 1996: 237)

[5] The analogy of the subject matter of ethics with a country does not, of course, hold throughout. While countries do not overlap, the subject matter of ethics overlaps with the subject matter of other philosophical branches.

The particular sentence Diamond refers to is the sentence '20 apples plus 30 apples is 50 apples', of which Wittgenstein says in his *Lectures on the Foundations of Mathematics* that it may be about apples as well as about numbers, depending on its use (Wittgenstein 1976: 113). Which category a sentence belongs to depends, as Diamond writes, 'not on what it is apparently about but on its use'. The word 'apparently' is of crucial importance here: a sentence may appear to be about something, but when we have a closer look at how it is used, it may turn out that we have been misled by the surface grammar of the sentence and that the sentence is actually about something else. So, although which category a sentence belongs to does not indeed depend on what it is *apparently* about, it *does* depend on what it is *actually* about. In Diamond's example, as in many other examples, we cannot decide on what the sentence is about or on what its subject matter is without looking at how the sentence is used. What the example shows is that subject matter is determined by use. A classification of the sentence '20 apples plus 30 apples is 50 apples' as a pomological rather than a mathematical sentence is a classification by subject matter, and because such a classification crucially depends on the use of the sentence, it can at the same time rightfully be conceived as a classification by use. Because a classification by use can at the same time be a classification by subject matter, showing that something is a classification by use does not amount to showing that it is *not* a classification by subject matter. When Diamond asks 'whether a sentence's belonging to ethics is a classification by use rather than by subject matter' (Diamond 1996: 237), she suggests that there is a conflict or opposition between use and subject matter, while 'Use or subject matter?' is, in the light of Wittgenstein's famous 'meaning is use' (Wittgenstein 2009: §43), a false dilemma.[6]

I do not want to suggest that there is no *distinction* to make between classifications by use and classifications by subject matter, I only want to stress that there is not necessarily a *conflict*, that the two types of classification are not mutually exclusive. As Diamond rightly points out, a classification of sentences as 'proverbs' is a classification by use and not by subject matter, while a classification as 'botanical' is a classification by subject matter. What our authors do not take sufficiently into account, however, is that the latter is *also* a classification by use. *Being a classification by use does not exclude being a classification by subject matter,* although certain classifications by use ('proverb') are not classifications by subject matter. Hence, even if the classification of sentences as 'ethical' is a classification by use (which is plausible), this has no direct consequences for the question of whether ethics has a subject matter, because there is no opposition between use and subject matter. Unfortunately, our authors need such a conflict (and I take the quotations at the beginning of this section to suggest such a conflict) if they want to argue from

[6] 'Meaning is use' is one of the most discussed topics in the literature about Wittgenstein, so I can hardly hope to offer an uncontroversial reading here. Nevertheless, I think that my reading is relatively innocent. While some authors insist on the strong view that, according to Wittgenstein, meaning is *identical* with use, I rely here only on the much weaker view that, according to Wittgenstein, there is no meaning without use, that meaning *depends on* use or is *determined by* use. Those who hold the strong view are also committed to the weaker one. For discussion of 'meaning is use', see Glock (1996).

the fact that 'ethical' is a classification by use to the conclusion that ethics has no particular subject matter.

There is a further problem with the proposal to conceive of 'ethical' as a categorization by use and not by subject matter. If ethical sentences are not ethical in virtue of their subject matter, but in virtue of their use, then one avoids questions about how to demarcate or characterize an ethical subject matter, but one immediately invites analogous questions about moral use: what is 'moral use'? How can it be demarcated or characterized? What kinds of uses of words and sentences are there and how is a moral use different from and similar to other kinds of uses? One could say: 'These questions need not be answered, because no further characterization, specification or demarcation of "the moral use" is possible or even desirable. Moreover, there is not just one "moral use", but there are many ways of using words and sentences in an ethical way.' Although this answer might be acceptable in other contexts, it cannot be accepted here, because analogous answers to questions about subject matter have been taken to make the very idea of an ethical subject matter problematic. An important part of the motivation to attack the idea of there being a subject matter and to claim that there can be no meta-ethics was the impossibility of compiling a closed list of distinctively moral concepts, to demarcate a particular subject matter or separable branch, to make (sharp) distinctions. May one not ask, then, how a moral use can be (sharply) distinguished or demarcated? Why would an indefinite variety of uses be more desirable than an overwhelmingly varied or complicated moral arena? 'Use' seems to invite exactly the same problems as 'subject matter'. The only significant difference is that philosophers have spent a great deal of effort trying to characterize the subject matter of ethics, and that we have acceptable demarcations and characterizations of the subject matter of moral philosophy (such as Moore's/Wittgenstein's that the subject matter of ethics is what is good), while demarcations or characterizations of the moral use(s) (as opposed to subject matter) are scarce. The only satisfying characterizations of 'moral use' one can give, I think, are those referring to a subject matter: a sentence or word is used in a moral way if it refers to what is good, or to what is absolutely good, or to what is intrinsically valuable, etc. This is the way in which I have understood 'moral use' when I said that a moral vocabulary would contain only words in their moral uses. Again, it is not wrong or misleading to talk about 'moral uses', but it is misleading to make it seem as if there were a conflict between use and subject matter, as if subject matter had nothing to do with use and *vice versa*.

Our authors' discontentment with 'subject matter' goes together, as we have seen, with a fear of moral philosophy becoming empty. If ethical sentences can be about anything, then the subject matter of moral philosophy would contain almost everything, and the idea of a moral philosophy becomes empty. My point is: if we characterize a subject matter, then ethical sentences cannot be about just anything. This, as we have shown, does not amount to setting limits to the ways in which moral thought might be expressed or fixing in advance what might count as ethical language, something to which Diamond (1996: 249), Richter (1996: 253) and Mulhall (2002: 310) rightly object. It does amount, however, to setting limits to what moral thought or moral sentences can be about, to saying, for example, that they are about 'what is good'. Characterizing or demarcating a subject matter thus prevents moral philosophy from becoming empty rather than causing it to be so. By contrast, refusing to characterize or demarcate what a moral use is *does* threaten to make 'moral use', and a conception of moral philosophy depending on it, empty, rather than preventing it from becoming so. Consider in this regard Hertzberg's saying that 'moral considerations *may* come to be expressed in *any* kind of thing we say' (Hertzberg 2002: 255). Thus, in meeting one of the main concerns of its proponents, a fear of emptiness, 'use' seems to come out worse than 'subject matter'. Rather than saving moral philosophy, an exclusive focus on use makes it impossible. Only Richter has seen that this is where the Diamond-Mulhall-Richter-Hertzberg-Conant argument may lead us, and he has accepted that consequence. Unfortunately, he attributes it also to Wittgenstein: 'It appears, then, that Wittgenstein was right to believe that there is no special ethical arena that could be the subject of moral philosophy or ethics. It does not follow from anything in the *Investigations* that moral philosophy is possible' (Richter 1996: 254).

Richter is mistaken in attributing his conclusion to Wittgenstein. First, as we have seen, Richter takes Wittgenstein's thinking about ethics to have changed little during his career. In 'A Lecture on Ethics', Wittgenstein explicitly says that ethics has a subject matter. Moreover, he speaks of 'the characteristic features of ethics' (Wittgenstein 2014: 43). Would this not be very strange if he took ethics to be impossible? Second, even if we assume that the later Wittgenstein changed his mind, Richter has not made clear why the later Wittgenstein would have believed that there is no special ethical arena. He may be right in saying that 'it does not follow from anything in the *Investigations* that moral philosophy is possible', but this is a strange way of putting things. Richter takes Wittgenstein to believe that moral philosophy is impossible, and one may thus expect him to point out that it *does* follow from something in the *Investigations* that moral philosophy is *impossible.* (Does it follow from anything in the *Investigations* that it is possible for me to go to China tomorrow? If not, then would one have to attribute to Wittgenstein the idea that it is impossible for me to do that?) Unfortunately, I cannot find a reference in Richter's article to something in the *Investigations* that would make moral philosophy impossible.

I conclude that Diamond *et al.* fail to make clear why a characterization of ethical sentences by use would avoid the problems they had with a characterization by subject matter. It seems only to invite more problems, especially because no convincing characterization by use of ethical sentences has been provided. That does not mean that an emphasis on use is pointless or irrelevant. By contrast, I think that Diamond *et al.* are right to insist that, in studying ethics, we should pay close attention to how sentences are used. Why? Because only if we attend to their use can we know what they are about, what their subject matter is, and whether and why they are ethical sentences.

4.5 'Ethics Is a Pervasive Dimension of Life Rather than a Distinguishable Region or Strand of It'

As an alternative to a characterization of the moral by subject matter, a characterization by use is problematic. Another alternative to a characterization by subject matter is to think about ethics as 'a pervasive dimension of human thought and action' (Conant 2002: 87), as 'a pervasive dimension of life rather than a distinguishable region or strand of it' (Mulhall 2002: 304). (Again, one may ask why Mulhall adds 'distinguishable'.) According to Mulhall, 'there is good reason to think that, from the beginning to the end of his philosophical career, Wittgenstein conceived of ethics in precisely this way' (Mulhall 2002: 304). As we have seen, there is good reason to question Mulhall's claim here, especially because he explicitly stresses the continuity in Wittgenstein's thinking on ethics. Still, thinking of ethics as a pervasive dimension may be an interesting option for moral philosophy. Let us, therefore, have a look at where it leads us.

The ethical as a pervasive dimension is meant to be contrasted with the ethical as a region, a strand, an area or an arena, in short, with a characterization of the ethical by subject matter. However, it cannot be straightforwardly equated with a characterization by use: the pervasive moral dimension is 'an ethical or spiritual aspect that is not retractable even in principle to certain kinds of words, or certain kinds of *uses* of words' (Mulhall 2002: 315, my italics; see also Hertzberg 2002: 255: 'not a part, but an aspect'). Diamond characterizes the ethical as a 'spirit, an attitude to the world and life', and this spirit 'can penetrate any thought or talk' (Diamond 2000: 153). According to Mulhall, it penetrates at least Wittgenstein's thought and talk: he asks us 'to consider the possibility that he [Wittgenstein] took [...] every one of his philosophical remarks – to have an ethical point' (Mulhall 2002: 321). This may all be true, but what is this ethical point and why is it ethical? So far, one could also say, for example, that the political, the aesthetical, the religious, the economical and even the epistemic are aspects rather than regions, not retractable to certain kinds of (uses) of words, attitudes to the world and life, spirits which can penetrate any thought or talk. Why not consider the possibility that Wittgenstein took every one of his philosophical remarks to have an aesthetical, a religious or a political point? I will ask of Mulhall what he asks of others: what is *distinctive* about the *moral* dimension? And why is it so pervasive in Wittgenstein's work? Mulhall and Diamond are not very clear on these matters (at least not in the articles under discussion), but the idea of a moral dimension can be found in other writings on Wittgenstein as well. I will discuss two ways in which it has been worked out. First, it has been understood in terms of a reconnection with the ordinary. Second, it has been connected to our learning to think better.

In her interesting essay 'A Critical Note on the New Mythology of the Ordinary', Marilena Andronico remarks: 'There is nowadays a widespread tendency to emphasize the ethical tone of Wittgenstein's philosophical work and to identify it with what is taken to be the rediscovery and acceptance of "the ordinary"' (Andronico

2013: 14).[7] Andronico traces the tendency back to Stanley Cavell, who points at the 'pervasiveness of something that may express itself as a moral or religious demand in the *Investigations*' and adds that 'the demand is not the subject of a separate study within it, call it "Ethics"' (Cavell 1988: 40). According to Andronico, the moral demand in the *Investigations* is often taken to be exemplified in Wittgenstein's controversial remark that 'What *we* do is to bring words back from their metaphysical to their everyday use' (Wittgenstein 2009: §116). This task, which is a task for philosophers, is then 'reinterpreted by Cavell and his disciples as, first and foremost, a *morally valuable* task, connecting us with something that can be called "our nature"' (Andronico 2013: 15). Andronico questions (rightly, I think) whether, and if so how, 'the recognition of ordinary uses of certain linguistic expressions can make the recognizer a morally better person' (Andronico 2013: 15). Even if we assume, for the sake of the argument, that it can, we are left with two problems for the 'moral dimension' account of the ethical. First, if the moral dimension is characterized in terms of rediscovery and acceptance of the ordinary, its pervasiveness in Wittgenstein's work can probably be accounted for, but why would it be a pervasive dimension of life? And if it is, why should we call it 'moral'? It seems to include many things we ordinarily would not call 'moral', and to exclude many things we ordinarily do call 'moral', hence to call for a radical revision of every ordinary conception of the ethical. (If the morally valuable lies in a rediscovery and acceptance of the ordinary, this might be a morally reprehensible undertaking.) Second, a characterization of the moral in terms of rediscovery and acceptance of the ordinary seems to reintroduce a subject matter: ethics is no longer about what is good, but about what is ordinary.

Another, related, interpretation of the moral dimension has been put forward by Conant. In 'On Going the Bloody *Hard* Way in Philosophy' (2002), he writes:

> When Wittgenstein writes '*Call me a truth-seeker and I will be satisfied*', he specifies the character of his striving in terms of something which is for him equally a philosophical and an ethical ideal. All philosophical thinking and writing accordingly has, for Wittgenstein, its ethical aspect. Wittgenstein thought that what (and more importantly *how*) we think is revelatory of who we are (and how we live), and that learning to think better (and, above all, to *change* the ways in which one thinks) is an important means to becoming a better – i.e. to becoming (what Wittgenstein calls) 'a real' – 'human being'. So, even though Wittgenstein, in one sense, 'has no ethics' (if 'ethics' names a branch of philosophy with its own proprietary subject matter), in another sense, his thinking and writing – on every page of his work – takes place under the pressure of an ethical demand. (Conant 2002: 90)

Again, we see that the 'ethical aspect' is identified with a philosophical task: learning to think better, philosophizing better (philosophizing as Wittgenstein wanted us to), is an important means to becoming a better person. If this is indeed what Wittgenstein thought, and if its pervasiveness in his work can be shown, and if this is the moral dimension pervading our lives (to be sure, Conant does not claim that, he is only concerned with the ethical demand of Wittgenstein's work), there is no

[7] The tendency is exemplified, according to Andronico, in Backström (2011), Conant (2002) and Gargani (2008).

way of contrasting this dimension with a subject matter. If our ethical task can be reduced to the task of thinking better, then what Wittgenstein proposes is that we think of the subject matter of ethics as 'what is good thinking' instead of 'what is good'. If it cannot be so reduced, learning to think better is only a means to becoming better human beings. The ultimate ethical task is thus our becoming better persons, which is perfectly in line with saying that 'what is good' or 'what it is to be a good person' is the subject matter of ethics. Characterizing the moral dimension in this way helps to make clear what distinguishes the moral dimension from other dimensions: becoming a better person is an ethical demand and not an aesthetical one, because it is about what is good and not about what is beautiful. I do not object to the idea of a moral dimension as such, but I suspect that, however one characterizes, specifies or demarcates the moral dimension or the ethical aspect, reference to a subject matter, to what ethics is about, will be necessary. Mulhall illustrates this. He writes that 'any and every Wittgensteinian philosophical exercise will place rigorous ethical demands upon its practitioners. For it will require them to acknowledge and respect the otherness of one's other' (Mulhall 2002: 319).[8] Why not say, then, that ethics is about acknowledgement and respect for the other's otherness and that this is (part of) the subject matter of moral philosophy? Why not say that the moral dimension, rather than being a radical alternative to the idea of there being a subject matter of moral philosophy, *is* the subject matter of moral philosophy?

An objection to this would be: a subject matter, like an area or region, has boundaries. To say that moral philosophy has a subject matter is to invite questions like 'Where are the boundaries?' and 'How can it be demarcated?' It seems impossible to answer these questions. By contrast, if the ethical is a pervasive dimension, there are no boundaries, and we do not have to answer questions about boundaries and demarcation. The problem with this objection is that it poses a false dilemma: either we should be able to say where the boundaries of moral philosophy are and how its subject matter can be demarcated, or there are no boundaries and there is no demarcation. But what if there were vague or blurred boundaries? What if there were a grey zone?[9] The dilemma does not have to be accepted, but if one accepts it, it is

[8] B.R. Tilghman provides a similar characterization of the moral dimension: 'It is a continuous reminder of the language, along with the underlying form of life, that constitutes human relationships, which are, perforce, moral relationships. [...] He [Wittgenstein] has reminded us what it is to discern the *humanity* in a man' (Tilghman 1987: 116–117). What I say about Mulhall's remark equally applies to Tilghman's.

[9] I agree with Lovibond in this respect. She provides a convincing characterization of the ethical and adds that 'we must resign ourselves to a state of affairs in which there will not always be a definite answer to the question whether this or that consideration is an "ethical" (as opposed to, say, to an "aesthetic" or an "educational") one. But though this complication may create a penumbra around the edges of the domain of ethical value, there is still a central area within which certain evaluative concepts (or in linguistic terms, predicates) will clearly fall' (Lovibond 2002: 33–34). Lovibond characterizes the ethical as follows: 'The ethical, let us say, pertains to what people learn to value through immersion in a community acquainted with ideas of right, duty, justice, solidarity, and common social or cultural interests extending beyond the lifetime of the present generation; and the transmission of any qualities regarded by the relevant community as virtues of character would no doubt be one "interest" figuring under the last of these headings' (Lovibond 2002: 33).

understandable that our authors reject its first horn. What they are left with is the second one, which easily leads to some obscure form of mysticism, in which even to say that the moral is a 'dimension' or an 'aspect' may be thought to set limits and to say too much. Again, Richter is the only one who follows the argument through:

> Ethics, for Wittgenstein, is not a subject nor a particular sphere or aspect of life. Everything is to do with ethics because life itself, the whole world, is, so to speak, a gift from God. (Richter 1996: 251)

Richter accepts this conclusion and takes it to be Wittgenstein's. It is not: Wittgenstein explicitly writes that 'My subject [...] is ethics' (Wittgenstein 2014: 43). Nevertheless, Richter understands that, if no subject matter of moral philosophy is characterized, no 'moral use' or 'moral dimension' specified (and how else could they be specified but by reference to a subject matter?), there is no objection to saying that 'everything is to do with ethics'. One of the motivations for objecting to the idea of a subject matter was that it allegedly threatens to make moral philosophy empty. We should not allow our 'notion of what counts as ethical' to 'become capacious to the point of emptiness' (Mulhall 2002: 303). I leave it to the reader to judge whether 'everything is to do with ethics' helps to take the fear of emptiness away.

4.6 Conclusion

I conclude that Wittgensteinian moral philosophy does not exclude the possibility of there being a particular subject matter of moral philosophy. There is room, within a Wittgensteinian framework, for understanding meta-ethics as it is commonly understood (unless one takes the common understanding to involve a demand for normative neutrality): as the second-order inquiry into what people are doing when thinking about and discussing what is intrinsically good, valuable, etc. That does not mean, of course, that Wittgensteinian moral philosophy is not different from traditional moral philosophy, but only that what distinguishes Wittgensteinian moral philosophy from traditional moral philosophy is not the absence of a subject matter or the impossibility of meta-ethics. What distinguishes Wittgensteinian moral philosophy (that is, what distinguishes its thinking about the subject matter of ethics from that of other moral philosophers) will depend, more than the authors I have discussed are prepared to acknowledge, on which Wittgenstein, early or late, one takes as its model. While I do not think that moral philosophy has very much to gain from Tractarian thought (which is not to say that there is nothing to gain), the *Investigations*, and the later Wittgenstein in general, have much more to offer. What distinguishes later Wittgensteinian moral philosophy from both the *Tractatus* and traditional moral philosophy is, roughly speaking, that it promises to be less one-sided. In focusing on certain aspects of its subject matter and thinking about it in certain ways, traditional moral philosophy has imprisoned itself in a small room of keywords and methods, and it has forgotten that there are alternatives, because the force of habit has made it overlook these alternatives again and again, up to the point of its not being able to see them anymore. Wittgensteinian moral philosophy *reminds* traditional moral philosophy of these alternatives: 'The work of the philosopher consists in marshaling recollections for a particular purpose' (Wittgenstein 2009: §127). It *draws attention* to them and *emphasizes* what has been overlooked in order for it to become visible again. The door of the room in which traditional moral philosophers have imprisoned themselves 'is unlocked, opens inwards', but it has not occurred to them 'to *pull*, rather than push against it' (Wittgenstein 1998: 48).

What are these alternatives? The value of some of the articles discussed here lies in their answers to *this* question (and not in their answers to questions about subject matter and meta-ethics). Hence, I am happy to refer to them for elaborated answers, and I will only point, by way of examples, at some of their general points of attention here. While traditional moral philosophy tends to focus on generalization, theory, rules, principles and judgments applicable across different contexts and circumstances in similar ways by different persons, later Wittgensteinian moral philosophy emphasizes the particularity and historicity of contexts and circumstances (forms of life and language-games). While traditional moral philosophy tends to focus on answers, Wittgensteinian moral philosophy focuses on questions and assumptions. Against moral dilemmas, the Wittgensteinian sets the creative capacity to reconceive and redescribe the situation so as to dissolve the dilemma. Against a focus on certain words and surface grammar: a focus on use. Against the search for commonalities and similarities between moral sentences and judgments: a focus on differences and the variety of the moral landscape.[10] Diamond *et al.* have succeeded, in (what I see no objection to calling) their meta-ethical work, in showing these to be distinguishing marks of Wittgensteinian moral philosophy. None of these marks, however, makes the idea of a subject matter for moral philosophy or the possibility of meta-ethics any less plausible.[11]

References

Andronico, Marilena. 2013. A Critical Note on the New Mythology of the Ordinary. In Mind, Language and Action. *Papers of the 36th International Wittgenstein Symposium*, ed. Danièle Moyal-Sharrock, Volker Munz, and Annalisa Coliva, 14–16. Kirchberg am Wechsel: Austrian Ludwig Wittgenstein Society.

Backström, Joel. 2011. Wittgenstein and the Moral Dimension of Philosophical Problems. In *The Oxford Handbook of Wittgenstein*, ed. Oskari Kuusela and Marie McGinn, 729–752. Oxford: Oxford University Press.

Cavell, Stanley. 1988. *This New Yet Unapproachable America. Lectures after Emerson After Wittgenstein*. Chicago: University of Chicago Press.

[10] On questions and answers in Wittgensteinian moral philosophy, see De Mesel (2014) and Chap. 7 of this book. Concerning the focus on differences rather than similarities between moral judgments, see Chap. 3, Sect. 3.5, and my forthcoming contribution in Kuusela and De Mesel (2018).

[11] I am grateful to Sabina Lovibond, Edward Harcourt, Stefan Rummens and the anonymous reviewers for their comments.

Conant, James. 2002. On going the Bloody *Hard* Way in Philosophy. In *The Possibilities of Sense*, ed. John H. Whittaker, 85–129. Basingstoke: Palgrave.

De Mesel, Benjamin. 2014. Moral Modesty, Moral Judgment and Moral Advice. A Wittgensteinian Approach. *International Journal of Philosophy and Theology* 75: 20–37.

———. 2017. Review of Edoardo Zamuner, Ermelinda Valentina Di Lascio, and D.K. Levy, eds. 2014. *Lecture on Ethics*. Oxford: Wiley-Blackwell. *Journal of Moral Philosophy* 14: 353–356.

Diamond, Cora. 1996. Wittgenstein, Mathematics, and Ethics. Resisting the Attractions of Realism. In *The Cambridge Companion to Wittgenstein*, ed. Hans Sluga and David G. Stern, 226–260. Cambridge: Cambridge University Press.

———. 2000. Ethics, Imagination and the Method of Wittgenstein's *Tractatus*. In *The New Wittgenstein*, ed. Alice Crary and Rupert Read, 149–173. London: Routledge.

Gargani, Aldo. 2008. *Wittgenstein. Musica, Parola, Gesto*. Milan: Raffaello Cortina Editore.

Glock, Hans-Johann. 1996. Abusing Use. *Dialectica* 50: 205–223.

Hertzberg, Lars. 2002. Moral Escapism and Applied Ethics. *Philosophical Papers* 31: 251–270.

Johnston, Paul. 1989. *Wittgenstein and Moral Philosophy*. London: Routledge.

Kuusela, Oskari, and Benjamin De Mesel, eds. 2018. *Ethics in the Wake of Wittgenstein*. New York: Routledge (forthcoming).

Lovibond, Sabina. 1983. *Realism and Imagination in Ethics*. Oxford: Blackwell.

———. 2002. *Ethical Formation*. Cambridge, MA: Harvard University Press.

Mulhall, Stephen. 2002. Ethics in the Light of Wittgenstein. *Philosophical Papers* 31: 293–321.

Richter, Duncan. 1996. Nothing to Be Said. Wittgenstein and Wittgensteinian Ethics. *The Southern Journal of Philosophy* 34: 243–256.

Sayre-McCord, Geoffrey. 2012. Metaethics. In *The Stanford Encyclopedia of Philosophy*, ed. Edward Zalta. Spring 2012 Edition. http://plato.stanford.edu/archives/spr2012/entries/metaethics/.

Schroeder, Severin. 2006. *Wittgenstein. The Way Out of the Fly-Bottle*. Cambridge: Polity.

Tilghman, B.R. 1987. The Moral Dimension of the *Philosophical Investigations*. *Philosophical Investigations* 10: 99–117.

Wittgenstein, Ludwig. 1976. In *Wittgenstein's Lectures on the Foundations of Mathematics*, ed. Cora Diamond. Chicago: University of Chicago Press.

———. 1998. *Culture and Value* . Rev 2nd ed. Oxford: Blackwell.

———. 2009. *Philosophical Investigations*. 4th ed., Trans. G.E.M. Anscombe, P.M.S Hacker, and Joachim Schulte. Oxford: Wiley-Blackwell.

———. 2014. A Lecture on Ethics. In *Lecture on Ethics*, ed. Edoardo Zamuner, Ermelinda Valentina Di Lascio, and D.K. Levy, 42–51. Oxford: Wiley-Blackwell.

Zamuner, Edoardo, Ermelinda Valentina Di Lascio, and D.K. Levy, eds. 2014. *Lecture on Ethics*. Oxford: Wiley-Blackwell.

Chapter 5
Wittgenstein and Objectivity in Ethics. A Reply to Brandhorst

In 'Correspondence to Reality in Ethics', Mario Brandhorst examines the view of ethics that Wittgenstein took in his later years. According to Brandhorst, Wittgenstein leaves room for truth and falsity, facts, correspondence and reality in ethics. Wittgenstein's target, argues Brandhorst, is objectivity. I argue (1) that Brandhorst's arguments in favour of truth, facts, reality and correspondence in ethics invite similar arguments in favour of objectivity, (2) that Brandhorst does not recognize this because his conception of objectivity is distorted by a Platonist picture, and (3) that he misinterprets the passage which he takes to support a Wittgensteinian case against objectivity.

5.1 Introduction

In his interesting and thought-provoking article entitled 'Correspondence to Reality in Ethics', Mario Brandhorst examines the view of ethics that Wittgenstein took in his later years (Brandhorst 2015, hereafter referred to as B). According to Brandhorst, the later Wittgenstein did not wish to deny that there is room for truth and falsity in ethics, that there are ethical facts, and that an ethical reality corresponds to ethical judgments. Rather, he wanted to warn us against the temptation to understand 'truth', 'falsity', 'fact', 'reality' and 'correspondence' in the ways in which we understand these notions in the context of a physical theory. These ways have no counterpart in ethics, but that does not imply that it makes no sense to speak of 'truth', 'falsity', 'fact', 'reality' and 'correspondence' in ethics. Brandhorst aims to show how Wittgenstein's later thought allows us to make sense of them in an ethical context.

De Mesel Benjamin. 2017. Wittgenstein and Objectivity in Ethics. A Reply to Brandhorst. *Philosophical Investigations* 40: 40–63. Published by Wiley-Blackwell. See http://onlinelibrary.wiley.com/doi/10.1111/phin.12129/abstract.

B. De Mesel, *The Later Wittgenstein and Moral Philosophy*, Nordic Wittgenstein Studies 4, https://doi.org/10.1007/978-3-319-97619-8_5

In the following, I will assume that Brandhorst's arguments are sound and that his claims about how the later Wittgenstein thought about truth, facts, reality and correspondence in ethics are true. I will show, however, that these arguments and claims do not support Brandhorst's conclusion. In his conclusion, he says that claims to *objectivity* in ethics, rather than claims to truth, facts, reality or correspondence, are Wittgenstein's real target and that this is confirmed by a passage from Wittgenstein's conversations with Rush Rhees in 1945 (B 247–248). I will argue that the passage does not confirm this and that Brandhorst's arguments actually support a case *for* objectivity in Wittgensteinian ethics rather than a case *against* it. That is, arguments similar to those invoked by Brandhorst can be (and have been) used to defend the claim that, according to the later Wittgenstein, there is room for objectivity in ethics. If Brandhorst's arguments succeed, they will allow for (rather than prevent) talk of objectivity in ethics.

I will proceed as follows. First, I will present a summary of the considerations, reasons and arguments used by Brandhorst to defend the claim that there is room for truth, facts, correspondence and reality in the later Wittgenstein's ethical thought (Sect. 5.2). Second, I will show how these considerations, reasons and arguments can be and have been used to defend conceptions of objectivity in ethics (Sect. 5.3). Third, I will argue that Brandhorst takes these arguments to speak against, rather than for, objectivity because his understanding of objectivity is distorted by a Platonist picture that is very similar to the pictures that he explicitly warns against when he discusses truth, facts, reality and correspondence in ethics (Sect. 5.4). Fourth, I will show how Brandhorst misreads the passage from Wittgenstein's conversations with Rhees, which he takes to support his case against objectivity (Sect. 5.5).

5.2 Brandhorst's Arguments

Let me start with an outline of Brandhorst's arguments, reasons and considerations (for a full account and defence, I am happy to refer to Brandhorst's article). I will structure them in a different way than Brandhorst's, namely, in a way that will help me to clarify, in Sect. 5.3, how they may contribute to a case for, rather than against, objectivity. Brandhorst combines these arguments, reasons and considerations in order to support the claim that there is room for truth, facts, reality and correspondence in the later Wittgenstein's thought about ethics.

5.2.1 *In a Different Way, Different Kinds, in a Different Sense*

Brandhorst recognizes that we cannot speak of truth, facts, reality and correspondence in ethics in the way in which we speak of them in, for example, physics. His overarching argument is that this does not mean that we cannot speak of these things in ethics at all. 'Truth', etc. are used in different ways in different language games.[1] Ethical outlooks *can* be divided into true and false, but true and false 'in a different sense' (B 231).[2] We must distinguish between 'different kinds of correspondence' (B 228).[3] Brandhorst refers, in this respect, to something that Wittgenstein said to Rhees: 'The way in which some reality corresponds – or conflicts – with a physical theory has no counterpart here [in ethics]' (Rhees 1965: 24). Nevertheless, Brandhorst claims that, '*There is* a kind of correspondence in the case of ethics, too. The notion of correspondence applies, but it must not be confused with the *kind* of correspondence that we find in physics [...]' (B 236). Brandhorst further aims to explain *in what sense* we can speak of truth, facts, correspondence and reality in ethics. The 'important question' is how these notions are to be understood in ethics (B 241).

5.2.2 *No View from Nowhere*

There is 'no view from nowhere in fundamental ethical orientation' (B 231). Suppose that we ask which ethical outlook, a Christian one or a Nietzschean one, is right or true. Such a question can only be answered 'given our ethical outlook, in the light of our attitudes and commitments' (B 231), and 'there is no such thing as "the ethical truth" independently of our ethical outlook' (B 246). There is 'no hope of transcending our personal ethical outlook by means of an appeal to the idea of ethical truth' (B 246). There is no such thing as transcending our personal ethical standpoint or orientation, as leaving it behind in order to assess, in an ethically neutral way, which outlook is right or true. To say that, for example, a Christian ethics is true means adopting it, or expressing a particular ethical stance (B 233–234). It is an illusion to think that we can divide ethical outlooks into true and false without adopting such a stance. That does not mean, however, that there is no room for saying that a particular ethical outlook is true. It only means that saying that a particular outlook is true does not involve any reference to an impersonal, ethically neutral standard.

---

[1] In this respect, Brandhorst refers to Wittgenstein (1977: III-§311f.).

[2] See also B 233 ('*what it is* for such an expression to be true or false can be very different from case to case').

[3] See also B 234 ('in an entirely different way') and B 241 ('a different idea of "a correspondence" or "responsibility" to "a reality"').

5.2.3 *There Is a Language-Game in Which the Expression Has a Use*

According to Brandhorst,

> […] there is a language-game […] making use of the words 'right' and 'wrong', 'true' and 'false' […], and […] it allows us to pass judgement on a given piece of advice or an ethical outlook. This includes judgements concerning the choice between a 'Christian' or a 'Nietzschean' attitude to marriage […] So there is a use for 'true' and 'false' even with respect to the highly general question of whether the Christian or Nietzschean ethical outlook is right. (B 231)

There is 'good use' for an expression such as 'although I believe that so and so is good, I may be wrong' (B 233). Brandhorst adds that 'we find it natural, even inevitable, to think of answers to fundamental ethical questions as true and of others as false' (B 231). It would be misleading to *deny* that answers to ethical questions can be true or false or to say that *no* reality corresponds to ethical expressions, because that would make their use 'seem arbitrary in a way in which it is clearly not' (B 243).[4] We have a use for these words that is not arbitrary, but 'firmly woven into the tapestry of our lives' (B 243). To say that no reality corresponds to ethical expressions would make ethical language into 'a game played merely for entertainment' (B 244).

5.2.4 *A Deflationary Conception, No Metaphysical Depth*

Wittgenstein's conception of truth is deflationary: '*p* is true' means simply '*p*'. Such a deflationary conception does not 'by itself rule out the idea that ethical truth may involve something more substantial', but it leaves room for the idea that ethical truth may *not* involve something more substantial, because an ethical judgment may justly be called true or false even if nothing more substantial is involved (B 231). Talk of truth has no metaphysical depth (B 232).

5.2.5 *With a Certain Conception of X, We Also Obtain a Conception of a Related Notion Y*

Brandhorst defends a deflationary conception of truth and claims that 'with this conception of ethical truth, we also obtain a conception of ethical facts. Instead of saying that *p* is true, we can also say that it is a fact that *p*; the concepts "truth" and

[4] Brandhorst uses these words in a discussion of mathematical expressions. But as we shall see in Sect. 5.2.6, he thinks that his suggestions with respect to truth, etc. in mathematical expressions naturally extend to ethics.

"fact" were made for each other' (B 232). If we can understand 'truth' in a deflationary way, then we can understand a related notion such as 'fact' in a deflationary way as well. If talk of truth has no metaphysical depth, then talk of facts has no metaphysical depth either. This suggests 'a particular way of understanding the uses of "true" and "false", "right" and "wrong", and *related notions* [my italics] in the context of ethical language' (B 233–234).

5.2.6 The Analogy to Logic and Mathematics

Brandhorst sees important analogies between Wittgenstein's talk of truth, facts, correspondence and reality in logic and mathematics on the one hand and his understanding of these notions in ethics on the other hand. According to his view, Wittgenstein's suggestions regarding truth, etc. in logic and mathematics naturally extend to ethics (B 242). In logic and mathematics, as in ethics, it is 'not at all obvious *how* [...] language relates to reality, and to *what* reality it could be said to relate' (B 237–238). In both cases, we run into difficulties when we understand truth, etc. as we understand them in physics (B 241). Brandhorst shows how Wittgenstein objects to G.H. Hardy's Platonist understanding of mathematics, in which there is a close analogy between physics and mathematics. Physics and mathematics are far more different than Hardy thought. Again, that does not mean that we cannot talk about truth, facts, reality and correspondence in mathematics. According to Brandhorst, Wittgenstein gradually softened his initial resistance to the idea that there are logical or mathematical facts (B 233). According to Wittgenstein's later views, logical and mathematical expressions such as '2 + 2 = 4' 'can be used to express a proposition, and propositions can be either true or false' (B 233). Wittgenstein does not deny that logical and mathematical statements correspond to reality, but asks 'over and over again what "correspondence to reality" might mean in a logical or mathematical context' (B 238; see also B 241). It emerges that 'there is a sense in which it is perfectly true and important that something real corresponds to our logical and mathematical language. It also emerges, however, that this reality is not what a Platonist might expect' (B 241). Brandhorst explains how, according to Wittgenstein, a reality can be said to correspond to logical and mathematical expressions (in a non-trivial way, that is, not in a way in which we simply translate 'it is true' into 'a reality corresponds to'). A reality corresponds to logical and mathematical expressions not in the way that it corresponds to propositions like 'it rains', but rather in the way that a reality corresponds to words like 'two' and 'perhaps'. A reality corresponds to 'it rains' if that proposition is true and we can assert it – that is to say, if it rains. By contrast, to say that a reality corresponds to words like 'two' and 'perhaps' could simply mean 'that we have a use for them' (Wittgenstein 1976: 249) (B 243–244).

5.3 Arguments for Objectivity?

Based on the outline in the previous section, I will argue in this section that considerations, reasons and arguments similar to those that Brandhorst uses in order to make a case for truth, facts, reality and correspondence in Wittgenstein's thought about ethics can also be used and have been used (with or without explicit reference to Wittgenstein) to make a case for objectivity in ethics. The authors I will mention do not all defend the same conception of objectivity, and it is not my aim to defend or discuss these conceptions in depth. Rather, by referring to different accounts of objectivity that make use of Brandhorst's Wittgensteinian arguments, I want to show that it is plausible that these arguments are all compatible with (different conceptions of) objectivity in ethics, and that some of them clearly point in the direction of an argument for objectivity rather than against it.

5.3.1 In a Different Way, Different Kinds, in a Different Sense

We can acknowledge that we cannot speak of objectivity in ethics in the same way in which we speak of it in, for example, physics. But that does not mean that we cannot speak of objectivity in ethics at all. What it is for something to be objective can be very different from case to case, and we can distinguish between different kinds or ideas or senses of objectivity.[5] Why not say that there is a kind of objectivity in ethics and that this notion must not be confused with the kind of objectivity that we find in physics? The important question, then, will be how objectivity is to be understood in ethics.

5.3.2 No View from Nowhere

Objectivity, in ethics and elsewhere, need not imply that there is a view from nowhere. That is, there is no requirement that, in order to be objective in ethics, we have to (try to) see things independently of our ethical outlook, to transcend our personal standpoint, to abstract from our attitudes and commitments. Being objective does not amount to being ethically neutral. To say that something is objective in ethics may well be to express a particular ethical stance. It does not need to involve any reference to an impersonal, ethically neutral standard.

Such a conception of ethical objectivity is not far-fetched. It has been argued in other domains that objectivity is not threatened by the impossibility of a transcendent perspective, of abstracting from a personal standpoint, from personal reactions,

[5] See, for example, distinctions made in Fisher and Kirchin (2006: 14) and Kirchin (2012: 25).

feelings and responses. In the context of a discussion about pain and the contemporary philosophy of mind, Roger Teichmann argues that 'in many cases, the foundations of objectivity reside in those very reactions and responses that get too easily classified as "merely subjective"' (Teichmann 2001: 35). According to Raimond Gaita,

> [...] to say that it is a requirement of objectivity that we step back to the point where we see human beings *as merely amongst* the many possible entities that can think, feel, etc., and that a human perspective is only one (limited) epistemic vantage point inferior to 'the point of view of the universe', is a classical philosophical instance of sawing away the branch on which one is sitting. (Gaita 2004: 168)

Gaita criticizes the 'illegitimate insistence that objective thought must always be separable from feeling' (Gaita 2000: 89). According to him, an objective judgment is not one that is *uninformed* by feeling or personal attitudes or commitments or outlooks, but one which is *undistorted* by them (Gaita 2000: 89).[6] Christopher Leich and Steven Holtzman claim that Wittgenstein stresses 'the *anthropocentric* nature of even our most objective practices of understanding, such as mathematics' (Leich and Holtzman 1981: 21).[7] They explicitly state that 'the moral objectivist need not [...] violate Wittgenstein's insights. He can maintain that moral assessments do express personal attitudes, and hence stand in an internal relation to action, while maintaining that they make a further cognitive claim' (Leich and Holtzman 1981: 17). And Alice Crary, defending 'a conception of objectivity no longer governed by an abstraction requirement' (Crary 2007c: 308), maintains that our rejection of the idea of a view from nowhere 'has no implications for our claim to basic epistemic ideals like objectivity' (Crary 2007b: 4; see also Crary 2007c: 307). She presents and defends an account of moral concepts

> [...] as determining features of the world that, while they need to be understood in terms of our attitudes, are nevertheless fully objective [...] What thus emerges is that here moral concepts trace out objectively consistent patterns, not [...] in a neutrally available region of fact, but rather in the moral outlooks within which they function. (Crary 2007a: 37–38)[8]

She refers, in this context, to the work of Stanley Cavell, Cora Diamond, Sabina Lovibond, John McDowell, Iris Murdoch and David Wiggins.

---

[6] See also Ripstein (1993: 359): the problem for objectivity is not that moral judgments always *depend* in many ways on personal standpoints, characteristics, reactions, etc., but that they sometimes *inappropriately depend* on them.

[7] See also Wiggins (1998: 101): '[...] a matter that is anthropocentric may be either more objective or less objective, or (at the limit) *merely* subjective'. Sabina Lovibond (1983: 40) quotes David Pears (1971: 171), who ascribes to Wittgenstein the idea that 'objectivism, in its only tenable form, collapses into anthropocentrism'.

[8] See Crary (2007a: especially Chapters 1 and 2) for a full-blown defence of this point.

5.3.3 There Is a Language-Game in Which the Expression Has a Use

Is there any use for talk of objectivity in ethics? It is probably less pervasive, at least in ordinary language, than talk about truth, but that holds for 'correspondence' as well. It certainly allows us (if 'true' and 'right' do, as Brandhorst claims) to pass judgment on a given piece of advice or a certain ethical outlook. There seems to be good use for such expressions as 'although I believe that so and so is good, it may not be objectively good' or 'although it appears to be good, it may not be objectively/really good' (on this point, see my comments on Brandhorst's reading of Wittgenstein in Sect. 5.5).[9] Whether we find it natural or inevitable to use 'objective' in ethical contexts can be discussed, but the same holds for correspondence and reality. In any case, if it is misleading to deny that answers to ethical questions can be true or false or to say that no reality corresponds to ethical expressions, because that would make their use seem arbitrary in a way in which it is clearly not, it seems at least equally misleading to deny that there is objectivity in ethics or to say that ethics is subjective. To say that something is subjective may suggest that it is 'a matter of whim or arbitrary decision, discretionary or unimportant' (Baker and Hacker 2009: 338),[10] as if we could decide what is ethically good. Arthur Ripstein delimits objectivity negatively, 'in terms of the absence of inappropriate factors' and 'by its contrast with some particular account of bias' (Ripstein 1993: 359, 361). In the light of his deflationary conception of objectivity, to say that a judgment is subjective is to say that certain inappropriate factors have influenced it, or that it is biased. 'To call a judgment subjective is either to disparage it openly, or else to politely suggest its irrelevance to justification' (Ripstein 1993: 370). If, then, one attributes to Wittgenstein the idea that ethics is essentially subjective, as Brandhorst does, one can easily be taken to suggest that something inherently inappropriate or biased is involved in all ethical judgment-making. That is certainly not what Wittgenstein, early or late, said or wanted to say.

5.3.4 A Deflationary Conception, No Metaphysical Depth

A conception of objectivity in ethics can be deflationary in the sense that it leaves room for the idea that objectivity may not involve 'something more substantial'. Talk of objectivity need not have any metaphysical depth. Ripstein has developed a

[9] Leich and Holtzman claim that 'Objectivity, on one well-established use of the term, is located in the distinction between appearance and reality; to maintain that it is an objective matter whether or not a certain speaker's claim is true is, on this use, to maintain that there is a clear difference between the claim's merely seeming to be true to the speaker and its actually being true' (Leich and Holtzman 1981: 2).

[10] Wittgenstein seems to link the objective to what is not a matter of decision in Wittgenstein 1974: §§270–271.

deflationary account of objectivity, 'faithful to the contexts in which the concept of objectivity is useful' (Ripstein 1993: 356), and he claims that his is 'the only idea of objectivity that we need (Or arguably, that we ever really had anyway, except for loaded uses in partisan philosophical debates.)' (Ripstein 1993: 361). According to his account, objectivity is not a property and a belief is objective if it satisfies a metaphysically innocent independence condition: in order for a belief to be objective, the fact that I think it must not make it so (Ripstein 1993: 358, 360). According to Paul Johnston, 'the real objectivity question in ethics is whether we accept the claim that there are ways of acting that everyone should recognise as correct. Holding this belief, however, does not involve any specific claims about the nature of what exists' (Johnston 1999: 117; see, on this issue, Johnston 1989: Chapter 7, 1999: Chapter 5). Crary deplores the assumption that, if we attempted to accommodate objective and intrinsically practical features of the world, 'we could not help but take for granted a category of metaphysically odd (or unnatural qualities)' (Crary 2007c: 302–303). Andrew Fisher and Simon Kirchin write that talk of objectivity need not commit us 'to any one particular ontological account' (Fisher and Kirchin 2006: 14).

5.3.5 With a Certain Conception of X, We Also Obtain a Conception of a Related Notion Y

Although the idea of a related notion is pretty elastic, 'objectivity' seems to qualify in a rather uncontroversial way as a notion related to 'truth', 'fact', 'correspondence' and 'reality'. Diamond, for instance, clearly recognizes this. She quotes Wittgenstein, who says that the words 'subjective' and 'objective' 'point to a difference between language-games' (Wittgenstein 2009b: §340), and adds: 'But, we may wish to say, it is not simply a matter of a difference of language-games, but of whether there can be truth and falsity in a particular language-game' (Diamond 1991b: 238). According to Wiggins, 'a subject matter is objective if and only if enough of the questions that are posed within it admit of answers that are substantially true – simply and plainly true, that is' (Wiggins 2006b: 359). Fisher and Kirchin write that 'when we say that a given area of thought or discourse [...] is objective we might well mean that [...] judgements typically made within the area are truth-apt' (Fisher and Kirchin 2006: 14).[11] Kirchin characterizes a sense of objectivity in which something is objective 'if it consistently features in the experiences of typical, mature human beings and whose *reality* [my italics] is hard to ignore' (Kirchin 2012: 132). Another notion related to 'truth', 'reality', etc. is 'knowledge'. Brandhorst writes: '[...] if there are truths [and, as we have seen, Brandhorst thinks that there *are* ethical truths, if only in some particular sense], there are facts, and with truths and facts there can be doubt, belief and knowledge'

[11] For another link between objectivity, truth and correspondence, see Ripstein (1993: 360–361).

(B 233). So, according to Brandhorst's Wittgenstein, there can be ethical knowledge, and Wittgenstein himself seems to acknowledge the link between knowledge and objectivity: 'It would be correct to say: "I believe …" has subjective truth; "I know …" not' (Wittgenstein 1974: §179).[12]

Thus, a deflationary conception of objectivity without metaphysical depth is not only independently plausible in the light of Wittgenstein's later philosophy, but the need for such a conception becomes even more pressing when 'truth', 'fact', 'reality', 'knowledge' and 'correspondence' are already understood in a deflationary, non-metaphysical way. A deflationary understanding of the latter notions invites a similar understanding of objectivity. Hence, Brandhorst's arguments do not just leave room for (allow for, are compatible with) a Wittgensteinian conception of objectivity or make such a conception possible. They actually contribute to the case for such a conception.

5.3.6 The Analogy to Logic and Mathematics

If Wittgenstein's suggestions regarding truth, reality, etc. in logic and mathematics naturally extend to ethics, there is no reason to assume that things would have to be otherwise regarding objectivity. While we will certainly run into difficulties if we understand objectivity in mathematics as we understand it in physics, this should not prevent us from talking about objectivity in mathematics. Nowhere does Wittgenstein deny that there is objectivity in mathematics.[13] Rather, he shows that mathematics is not objective in the way that a Platonist might expect, because 'mastery of mathematics depends on our possession of a complex cluster of attitudes – senses of what is important, relevant, even senses of what constitutes a good reason for drawing a particular inference' (Leich and Holtzman 1981: 21). Wittgenstein's philosophy of mathematics brings out that mathematics and ethics are more similar in certain respects than we tend to think and suggests that objectivity is one of these respects.[14] Wiggins remarks that

[12] For another link between objectivity, truth and knowledge, see Lovibond (2002: 15–16).

[13] Although some commentators have read him as if he did. Crispin Wright (1980) is an example. For a good discussion of Wright's Wittgenstein and objectivity in logic and mathematics, see Diamond (1991a). Diamond shows that Wittgenstein's later philosophy of logic and mathematics is 'distorted if seen in anti-realist terms, and does not involve the denials of objectivity Wright sees' (Diamond 1991a: 217). Another example of a commentator who reads Wittgenstein as if he denied that there is objectivity in logic and mathematics is Michael Dummett (1959). Diamond discusses 'the impression, given to Dummett and others, that Wittgenstein denies the *objectivity* of proof. It looks as if Wittgenstein is saying that it is all really subjective, after all, only what characterizes it is that we *say* it is not. But Wittgenstein is not denying that there is all the difference between what is really an objectively valid proof and one which only appears valid' (Diamond 1991c: 256). See, in this respect, Wittgenstein 2009b: §348 and Baker and Hacker 2009: 297.

[14] Disagreement is another. See Leich and Holtzman (1981: 21) on Wittgenstein's famous example of the wood-sellers (Wittgenstein 1978: I-§143 f.).

> [...] for someone who wanted to combine objectivity [in ethics] with a doctrine of qualified cognitivism or of underdetermination, there might be no better model than Wittgenstein's normative conception of the objectivity of mathematics; and no better exemplar than Wittgenstein's extended description of how a continuing cumulative process of making or constructing can amount to the creation of a shared form of life that is constitutive of rationality itself [...]. (Wiggins 1998: 128; see also Wiggins 2006a: 330, and 2006b: 366–367)

Moreover, Wiggins believes that the resulting conception of objectivity in ethics will be harmonious with 'the rough and ready ordinary acceptation of the word' (Wiggins 2006b: 370).

In the light of Brandhorst's arguments and Wittgenstein's philosophy of mathematics, it would be misleading to deny that mathematics is objective, because it would make the use of mathematical expressions seem arbitrary in a way that they are clearly not. Brandhorst's suggestion to understand Wittgensteinian ethics according to the model of Wittgenstein's understanding of mathematics does not just make it possible to talk about objectivity in ethics in a way similar to the way in which it is possible to talk about objectivity in mathematics. It actually contributes to the case for objectivity in ethics because it is much less controversial to say that there is objectivity in mathematics than to say that there is objectivity in ethics. A Wittgensteinian approach to objectivity in mathematics and in ethics would involve asking over and over again what 'objectivity' might mean in mathematical and ethical contexts, and it would emerge that these kinds of objectivity are not what a Platonist might expect.

5.4 Brandhorst's Conception of Objectivity

If Brandhorst's own Wittgensteinian considerations, reasons and arguments in defence of truth, etc. in ethics allow, *prima facie* (that is, on the basis of my brief outline of these arguments), for objectivity in ethics and if they have indeed been frequently used to make a case for objectivity in ethics, then why does Brandhorst think that objectivity is Wittgenstein's real target?[15] I will argue that it is because his

[15] According to Brandhorst, another target of Wittgenstein is realism (B 21). In the light of Brandhorst's own arguments, this is a strange claim. Does he not ascribe a form of realism to Wittgenstein when he says that, according to Wittgenstein, a reality corresponds to ethical expressions? Brandhorst seems to think that we can only speak of realism when the reality that we appeal to is of a particular kind, more specifically the kind of reality that corresponds to expressions in physical theory. He says that realism is Wittgenstein's target because 'claims to truth and objectivity that are the hallmark of the realist perspective cannot be maintained' (B 21). In this chapter, I argue that claims to objectivity can be maintained, if only in a particular sense. But that just means that the realism we can ascribe to Wittgenstein in the light of Brandhorst's arguments will be a particular kind of realism, not that realism as such (rather than Platonist forms of it) was his target, as Brandhorst claims. With respect to truth, Brandhorst seems to think that realists cannot understand truth as Wittgenstein understands it. But that is not true. As Brandhorst himself acknowledges early in his article, a Wittgensteinian deflationary conception of truth 'does not by itself rule out the idea that there is an objective reality *of one kind or another* [my italics] to which a given

conception of objectivity in ethics is distorted by a Platonist picture that is similar to the picture that he explicitly warns about in talking about truth, facts, reality and correspondence in ethics.

(1) First, Brandhorst sees a necessary link between the objective on the one hand and the impersonal and the neutral on the other. When he discusses ethical truth, he rightly emphasizes that Wittgenstein does not leave room for the idea of a view from nowhere. Ethical questions can only be answered in the light of our ethical outlook, our attitudes and commitments. There is no such thing as transcending our personal ethical standpoint in order to assess, in an ethically neutral way, whether an ethical outlook or judgment is right or true. But Brandhorst does not just dismiss the idea of impersonal or neutral ethical truth. Instead, he dismisses 'the idea of objective, impersonal truth' (B 231). In his view, there cannot be room for objectivity in ethics because there is no room for impersonal and neutral ethical truth. That does not follow, however, if a conception of objectivity as outlined in Sect. 5.3.2 that is faithful to the contexts in which we use the concept can be developed.

Brandhorst concludes that, for Wittgenstein, 'ethics remains deeply personal. It is bound up with the deepest concerns and commitments of those who see the world around them from a singular, ethical point of view' (B 245–246). He adds that 'in this respect, there is a surprisingly deep continuity between the later work and the *Tractatus*. Regardless of its ties to a social world, ethics is essentially subjective. It is bound to a particular perspective, a point of view towards the world' (B 246). It is certainly true that Wittgenstein, both early and late, saw ethics as deeply personal and bound up with people's deepest concerns and commitments. It does not follow, though, that he saw ethics as essentially subjective. That only follows if what is deeply personal and bound up with people's deepest concerns and commitments cannot be objective.[16] It follows if one adheres to a conception of objectivity in which objectivity requires the possibility of a view from nowhere. It is remarkable that, while Brandhorst rightly objects to Platonist view-from-nowhere accounts of truth, facts, correspondence and reality, he implicitly commits himself to such a view-from-nowhere account of objectivity. He succumbs, in my view, to a temptation described by Wittgenstein in *The Blue Book:*

> When we look at everything that we know and can say about the world as resting upon personal experience, then what we know seems to lose a good deal of its value, reliability, and solidity. We are then inclined to say that it is all 'subjective' [...] This should remind us of the case when the popular scientist appeared to have shown us that the floor which we stand on is not really solid because it is made up of electrons. (Wittgenstein 1969: 48)

Just as the fact that the floor is made up of electrons should not, in Wittgenstein's view, prevent us from saying that it is solid, although popular scientists are inclined

true proposition corresponds' (B 6). A deflationary conception of truth is perfectly compatible with realism.

[16] And, one should add, if what is not objective is thereby subjective. That may seem obvious, and it is clearly obvious for Brandhorst, but see Wiggins (2006b: 377).

to say that it is not really solid, the fact that we cannot transcend our personal standpoint in ethics should not prevent us from saying that some ethical judgments are objective, although some ethicists are inclined to say that they are all subjective. Just as some popular scientists misunderstand what it means to say, in everyday language, that something is solid, and confuse it with the meaning of solidity in specialized, scientific accounts, Brandhorst seems to misunderstand what it means to say, in everyday ethical language, that something is objective, and confuses it with the meaning of objectivity in a Platonist view-from-nowhere account.

As Brandhorst notes in the beginning of his article, the later Wittgenstein 'abandoned his Tractarian conception of reality, language and representation. Without that conception, the claim that "ethics cannot be expressed" lacked its former rationale; and in the light of the later conception of language that slowly emerged, it is hard to see what a new rationale could be' (B 227–228). Brandhorst would admit that the conception of reality, language and representation outlined in the *Tractatus* is a view-from-nowhere account and that, with regard to objectivity too, we cannot expect anything but a view-from-nowhere account from the *Tractatus*. According to such an account, there are no ethical truths or ethical facts and there is no correspondence between ethical language and ethical reality. Similarly, according to such an account, there is no objectivity in ethics. We can agree with Brandhorst here and say that the early Wittgenstein saw ethics as essentially subjective.[17] But without a Tractarian view-from-nowhere account of objectivity, that claim lacks its former rationale. Why would Wittgenstein have abandoned his conceptions of reality, language and representation while keeping his conception of objectivity intact? Is a Tractarian conception of objectivity not as untenable as Tractarian conceptions of reality, language and representation, and for largely the same reasons? The later Wittgenstein gives us no reason to suppose that, in order to be objective, we have to abstract from our personal commitments and occupy some kind of Archimedean point of view, since there *is* no such point of view to occupy. Objectivity does not have to be understood as incompatible with the expression of personal commitments. The fact that we cannot abstract from the latter gives us no reason to claim that Wittgenstein still saw ethics as essentially subjective. Indeed, according to a view of objectivity purged from Tractarian misconceptions, Wittgenstein's rejection of the Hamletian thought that 'nothing is either good nor bad, but thinking makes it so' (Wittgenstein 2014: 45) could be enough to justify talk of objectivity in ethics (on this point, see the striking similarity between the formulations of Wittgenstein's rejection of the Hamletian thought and Ripstein's deflationary account of objectivity in Sect. 5.3.4). Thus, the continuity that Brandhorst sees between the later work

[17]Although it is probably more accurate, within a Tractarian framework, to say that it *makes no sense* to speak of ethical truths or facts or objectivity in ethics or correspondence of ethical language with an ethical reality. It does not follow from 'it makes no sense to say that ethics is objective' that 'ethics is subjective' or that 'ethics is essentially subjective'. What follows is that it makes no sense to say that ethics is subjective either. For the sake of the argument, however, I grant that Brandhorst is right in saying that, for the early Wittgenstein, ethics is subjective.

and the *Tractatus* is not as deep as he suggests. Ethics remains deeply personal, but the later Wittgenstein is not committed to the claim that it is essentially subjective.

(2) Brandhorst's conception of objectivity in ethics is distorted by a picture he warns against in talking about truth, etc. in a second way. He writes:

> So Wittgenstein also accepts the idea that an ethical judgement may justly be called true or false. Moreover, he accepts it in a sense that goes beyond a 'relative' sense that merely fits a given ethical judgement into a given ethical outlook, using standards of coherence. But as it stands, this concession is worthless for the objectivist. He wants – and needs – something more substantial than that. (B 232)

And: 'All the same, if there are truths, there are facts, and with truth and facts there can be doubt, belief and knowledge. But as before, these concessions have no metaphysical depth. So by themselves, they cannot give objectivists what they require' (B 233).[18] Apparently, the objectivist wants and needs 'something more substantial', something Brandhorst (and Wittgenstein) cannot give him: metaphysical depth. If the objectivist is simply someone who believes that there is objectivity in ethics, then this is a remarkable claim.[19] Brandhorst does not argue for this claim, he simply asserts that the objectivist needs something more substantial. But that is not true if a deflationary account of objectivity as outlined in Sect. 5.3.4, faithful to the contexts in which we use the concept, can be developed. Such a conception is invited by Brandhorst's own arguments, on behalf of Wittgenstein, for truth, facts, correspondence and reality in ethics. Brandhorst takes pains to argue that truth, etc. do not require metaphysical depth and that these notions will mislead us if we 'superimpose [...] Platonist pictures' (B 244). But he overlooks the fact that the same may be true, for largely similar reasons, of objectivity.

5.5 How Brandhorst Reads Wittgenstein on Objectivity

I have argued that Brandhorst's considerations, reasons and arguments invite, rather than resist, the idea of objectivity in Wittgensteinian ethics, but that Brandhorst cannot see this because he is wedded to a conception of objectivity of a Platonist kind,

---

[18] See also B 245: '[...] a "quality" as the objectivist conceives of it'.

[19] One could object here that perhaps Brandhorst sees the objectivist not simply as someone who believes that there is objectivity in ethics, but rather as someone who believes that there is a particular kind of objectivity in ethics, more specifically the kind of objectivity that we find in physics. In that case, the belief that there is objectivity in ethics would not make one an ethical objectivist, as the belief that ethical expressions correspond to a reality does not make one an ethical realist in Brandhorst's view (see footnote 15), and the objectivist would indeed require 'something more substantial'. However, if this were the reading that Brandhorst favours, then what he says about the objectivist could not support his case, on behalf of Wittgenstein, against claims to objectivity in ethics. At most, it could support a case against a kind of Platonist objectivism. However, Brandhorst makes it clear that he thinks that Wittgenstein is not just arguing against 'overly ambitious forms of Platonism', but that realism and objectivity are his targets, and that *that* is what he wishes to make plausible (B 21–22).

the kind of conception he explicitly warns against in talking about truth, facts, reality and correspondence in ethics. So far, I have ignored the fact that Brandhorst uses some quotations from Wittgenstein in order to support and confirm his position about objectivity in the later Wittgenstein's thought about ethics (B 248). I will argue in this section that, influenced by his view-from-nowhere conception of objectivity, he has misinterpreted these quotations.

I will start with a brief outline of how Brandhorst wants us to read Wittgenstein. His discussion of Wittgenstein's thought about truth, facts, correspondence and reality in ethics leans heavily on his reading of a remark, already quoted in Sect. 5.2.1, in which Wittgenstein says to Rhees that the 'way in which some reality corresponds – or conflicts – with a physical theory has no counterpart here [with respect to ethical judgment]'. Brandhorst comments on this remark as follows: 'When we read his words carefully, Wittgenstein does not seem to want to simply *deny* that there can be a correspondence between some ethical judgement and some reality' (B 236). He makes similar comments later on: '[…] it is not clear that we should simply say that *no* reality corresponds to "2 + 2 = 4"' (B 241); '[…] Wittgenstein does not wish to *deny* that "a reality corresponds" to logical or mathematical language' (B 244). And he supports these comments by a quotation from Wittgenstein's *Lectures on the Foundations of Mathematics:* 'I don't say: "No reality corresponds."' (Wittgenstein 1976: 249). He concludes that 'The right thing to say is precisely what Wittgenstein says: *The way in which* a reality corresponds – or conflicts – with a physical theory has no counterpart here. As before, we should not say: "No reality corresponds"; and as before, Wittgenstein makes a point of not saying it' (B 245).

In short, Brandhorst frequently warns us against the temptation to read Wittgenstein as if he denied something which he only questions or problematizes (see also B 235 and B 240, footnote 47). The temptation to read Wittgenstein in such ways has been noted, for instance, by Oswald Hanfling. Hanfling sees this temptation as a way of getting into difficulties with Wittgenstein. It is 'a craving to look for hard and fast claims where what one is given is tentative investigations; for proofs and refutations where "description alone" is intended; and for answers where only questions are given' (Hanfling 2002: 12). Brandhorst recognizes the danger of such cravings for hard and fast claims, proofs and refutations. He often uses italics to emphasize 'deny' and 'no'. In that way, he makes it clear that we should not ascribe to Wittgenstein the hard and fast claim that no reality corresponds to ethical language and that we should not take him to refute the idea of such a correspondence.

Brandhorst is right in reading Wittgenstein as he does in the context of correspondence and reality; that is, in not reading denials into Wittgenstein. He is less careful, however, and less true to his own hermeneutical maxims, in his treatment of two other passages.

(1) In his *Philosophical Investigations,* Wittgenstein says about objectivity in mathematics that 'what a mathematician is inclined to say about the objectivity and reality of mathematical facts is not a philosophy of mathematics, but something for philosophical *treatment*' (Wittgenstein 2009a: §254). Brandhorst quotes this in a footnote in order to illustrate that Wittgenstein's 'therapeutical

task is to uncover the various pictures, assumptions and fragments of theory that encourage the Platonist interpretation' (B 241). He does not notice, however, that Wittgenstein not only mentions what the mathematician is inclined to say about *reality* as a candidate for treatment, but also what he is inclined to say about *objectivity*. For Wittgenstein, objectivity and reality in mathematics very much seem to be in the same boat. One way to read this passage would be to say that Wittgenstein denies that mathematical facts can be real or objective. Such a reading, in which we take Wittgenstein to deny something which he merely questions or problematizes, goes against the way in which Brandhorst asks us to read Wittgenstein. Another way to read this passage, the most natural and right way in my view, and the way that aligns with Brandhorst's recommendations, would be to say that Wittgenstein does not *deny* that mathematical facts can be real or objective. He does not say that there is no reality or no objectivity in mathematics. He merely warns us against a certain way of understanding what reality and objectivity in a mathematical context amount to. If we understand reality and objectivity in the way that the Platonist and certain mathematicians understand them, then what we will say about the reality and objectivity of mathematical facts will make no sense, and philosophical treatment will show that. So if we read the passage in the way that Brandhorst recommends that we read Wittgenstein, and if Wittgenstein's suggestions with respect to truth, reality, etc. in mathematics naturally extend to ethics, as Brandhorst claims, we have every reason to suppose that reality and objectivity will be in the same boat there as well. But that is exactly what Brandhorst denies or is at least committed to denying. While he recognizes that we can speak of reality in Wittgensteinian ethics, claims to objectivity in ethics are said to be Wittgenstein's target. Thus, Brandhorst's conclusion about objectivity in Wittgensteinian ethics seems to be at odds with his preferred way of reading Wittgenstein.

(2) According to Brandhorst, a passage from Wittgenstein's conversations with Rhees confirms that claims to objectivity in ethics are Wittgenstein's target. Rhees reports:

> He [Wittgenstein] came back to this question of 'the right ethics' later. He did so once (in 1945) when he was discussing the relations of ethics and psychology and sociology. 'People have had the notion of an ethical theory – the idea of finding the true nature of goodness or of duty. Plato wanted to do this – to set ethical inquiry in the direction of finding the true nature of goodness – so as to achieve objectivity and avoid relativity. He thought relativity must be avoided at all costs, since it would destroy the imperative in morality.' (Rhees 1965: 23)

This passage from Wittgenstein's conversations with Rhees can only confirm that claims to objectivity in ethics are Wittgenstein's target if Wittgenstein is taken to *deny* that there is objectivity in ethics. But, *pace* Brandhorst, Wittgenstein does not deny that. What we have here is 'description alone', and the temptation or craving to look for hard and fast claims should be avoided. If Brandhorst would have read this passage as he read the passage about there being no counterpart in ethics for the way in which some reality corresponds with a physical theory, he would have emphasized that Wittgenstein does not say that there is *no* objectivity in ethics. If this passage about Plato and objectivity confirms that Wittgenstein's target was objectivity in ethics, then the passage about physical theory will confirm that Wittgenstein's target was reality in ethics. Brandhorst, however, uses the counterpart passage to defend the idea that there is room for correspondence to reality in ethics, and the passage about Plato to deny that there is room for objectivity in ethics.

If the passage about Plato should not be read as if Wittgenstein denied that there is objectivity in ethics, then how could it be read? Wittgenstein talks about what people and Plato wanted to do, namely, to find the true nature of goodness or of duty so as to achieve objectivity and avoid relativity. He does not say that this cannot be done or that there is no such thing as objectivity in ethics. Rather, in line with the passage about objectivity in mathematics and his warnings against Platonist pictures, I suggest that he could be read as saying that what people and Plato (and Platonists) are inclined to say about the true nature of goodness or duty, or what they are inclined to say about objectivity in ethics, is something for philosophical treatment. Thus, contrary to what Brandhorst claims, Wittgenstein's target in this passage is Platonism (and people inclined to Platonist talk) rather than objectivity. Once more, the 'important question' is how 'true nature' and 'objectivity' are to be understood.

Brandhorst comments on the passage as follows:

> Presumably, to find the 'true nature' of goodness or duty would be to find the criterion by which we could judge what is *really* good, as opposed to merely *apparently* good, or good *from our point of view*. In other words, that criterion would be objective. [...] That search was based on an illusion, and it is a further illusion to think that without objectivity, there can be no imperative in ethics. (B 22)

Brandhorst reads Wittgenstein as if he were saying that there is no place for objective criteria in ethics. He opposes what is really or objectively good to what is apparently good or good from our own point of view. He thinks that because we cannot abstract from our point of view in ethics and because there is no view from nowhere, things cannot be objectively good (remember, in this regard, the passage from *The Blue Book* in Sect. 5.4). But that only follows if one adheres to a conception of objectivity as outlined in Sect. 5.4, and not if a Wittgensteinian conception of objectivity as outlined in Sect. 5.3.2, which is more faithful to the contexts in which we use the concept, can be developed. Neither the search for the true nature of goodness or duty nor the search for objective criteria in ethics are based on illusions, but the idea that objectivity necessarily excludes our personal point of view.

Brandhorst's comment is misleading, moreover, in that it suggests that it does not make sense to talk about 'really good' as opposed to 'apparently good'. But there is good reason for such a contrast. Brandhorst recognizes that there is good reason to use the expression 'although I believe that so and so is good, I may be wrong' (see also Sect. 5.2.3), and he adds that this expression should not be taken to involve 'a commitment to ethical truth that transcends any personal standpoint' (B 233). Why not say, then, that there is a place for an expression such as 'although it appears good to me, it may not be really/objectively good', and add that this expression should not be taken to involve a commitment to ethical reality or objectivity that transcends any personal standpoint? It does not make sense to speak of things being really or objectively good if that means 'good from a God's eye point of view', but that does not mean that things or deeds cannot be really good as opposed to being merely apparently so.

Or does it? One could defend Brandhorst as follows. If there is no view from nowhere, if we cannot transcend our personal standpoint in ethics, then 'it is good' can only mean 'it is good from my point of view'. And is that not exactly what 'it appears to be good' means? Does the distinction between 'it is good' and 'it appears to be good' not collapse when it becomes a distinction between 'it is good from my point of view' and 'it appears to be good'? The answer is no. An analogy may help to illustrate this. Suppose that we are asked what colour something is and that we are not allowed to use scientific instruments in order to measure wavelengths, etc. We cannot occupy a view from nowhere in order to look at the object; we only have our own personal standpoint. We say that the object is red. Here, 'it is red' is not equivalent to 'it appears to be red', because the latter expression suggests that we are not sure. We would use it, for example, when it is dark, when we think that our eyes may not be functioning well or when the distance between us and the object does not allow us to make out its colour; in short, when circumstances are such that we would not sincerely claim to *know* that it is red. If we know that something is red (and it seems obvious that we can justly be said to know that some things are red even if we cannot transcend our personal standpoint), we usually do not say that it appears red to us. The latter may not be wrong, but it would at least be misleading.[20] Similarly, if we (think that we) know that something is good, we usually do not say that it appears good to us, for that would be misleading too. As we have seen, Brandhorst recognizes that the later Wittgenstein's thought allows for ethical knowledge, and Wittgenstein links knowledge to objectivity. The ethical knowledge Wittgenstein allows for does not require a view from nowhere, because, as Brandhorst rightly stresses, there can be no such thing in Wittgensteinian ethics. It is knowledge 'from my point of view'. Now if there is such ethical knowledge and if 'appear' suggests (maybe not always, but at least sometimes) that we do not know, then the distinction between 'it is good' and 'it appears to be good' does not collapse: it will often be possible to read it as a distinction between 'I know (from my point of view) that it is good' and 'it appears (from my point of view) to be good, but I am not sure, I may be wrong, mistaken, etc.'

In summary, we can say that Brandhorst has misread the passage that he thinks confirms that claims to objectivity in ethics are Wittgenstein's target. Going against his own recommendations for reading Wittgenstein, he reads the passage as if Wittgenstein denied that there is a meaningful contrast between what is really good and what is apparently good, and as if he denied that there is objectivity in ethics.

---

[20] The analogy is meant to clarify a point about our use of 'appear', and nothing more. I do not want to suggest that the analogy between seeing that something is good and seeing that something is red holds throughout. See, on this point, Chap. 6.

The reason why Brandhorst misreads Wittgenstein lies in his conception of objectivity, a conception that links the possibility of objectivity to the possibility of a view from nowhere.

5.6 Conclusion

I conclude that Brandhorst has not provided convincing reasons or arguments for his conclusion that Wittgenstein's target is objectivity in ethics. Rather, his considerations, reasons and arguments in favour of truth, facts, correspondence and reality in the later Wittgenstein's thought about ethics speak for objectivity. Brandhorst has shown that he is sensitive to the demand of not reading refutations into Wittgenstein, but he has failed to extend that sensitivity to the case of objectivity. With respect to objectivity, Wittgenstein often problematizes and asks questions without answering them,[21] but he nowhere says or implies that there is *no* objectivity in ethics. That Wittgensteinian ethics is compatible with the idea of objectivity in ethics should not be taken to mean that *there is* objectivity in ethics according to Wittgenstein. Instead, it should only be taken to mean that that cannot be excluded if Brandhorst's arguments are sound and if his recommendations for reading Wittgenstein are accepted. Brandhorst faces a dilemma: either he has to give up on, or to qualify, the claim that Wittgenstein's target is objectivity in ethics, or he has to provide extra reasons for thinking that Wittgenstein took considerations and arguments similar to those offered in favour of truth, etc. to be inapplicable to objectivity.[22]

References

Baker, G.P., and P.M.S. Hacker. 2009. *Wittgenstein: Rules, Grammar and Necessity. Volume 2 of an Analytical Commentary on the Philosophical Investigations. Essays and Exegesis of §§185–242*. 2nd ed. extensively revised by P.M.S. Hacker. Oxford: Wiley-Blackwell.

Brandhorst, Mario. 2015. Correspondence to Reality in Ethics. *Philosophical Investigations* 38: 227–250.

———. 2017. Varieties of Objectivity. Reply to De Mesel. *Philosophical Investigations* 40: 64–81.

Crary, Alice. 2007a. *Beyond Moral Judgment*. Cambridge, MA: Harvard University Press.

———, ed. 2007b. *Wittgenstein and the Moral Life. Essays in Honor of Cora Diamond*. Cambridge, MA: MIT Press.

———. 2007c. Wittgenstein and Ethical Naturalism. In *Wittgenstein and his Interpreters. Essays in Memory of Gordon Baker*, ed. Guy Kahane, Edward Kanterian, and Oskari Kuusela, 295–319. Oxford: Wiley-Blackwell.

[21] See, for instance, 'But is there then no objective truth?' (Wittgenstein 1974: §108) and 'But is there no objective character here?' (Wittgenstein 1974: §336).

[22] This chapter is a reply to Brandhorst (2015). For Brandhorst's reply, see Brandhorst (2017). I am grateful to Stefan Rummens for his comments on an earlier version of this chapter.

Diamond, Cora. 1991a. Wright's Wittgenstein. In *The Realistic Spirit. Wittgenstein, Philosophy, and the Mind*, 205–223. Cambridge, MA: MIT Press.

———. 1991b. Secondary Sense. In *The Realistic Spirit. Wittgenstein, Philosophy, and the Mind*, 225–241. Cambridge, MA: MIT Press.

———. 1991c. The Face of Necessity. In *The Realistic Spirit. Wittgenstein, Philosophy, and the Mind*, 243–266. Cambridge, MA: MIT Press.

Dummett, Michael. 1959. Wittgenstein's Philosophy of Mathematics. *The Philosophical Review* 68: 324–348.

Fisher, Andrew, and Simon Kirchin, eds. 2006. *Arguing About Metaethics*. London: Routledge.

Gaita, Raimond. 2000. *A Common Humanity. Thinking About Love and Truth and Justice*. 2nd ed. London: Routledge.

———. 2004. *Good and Evil. An Absolute Conception*. 2nd ed. London: Routledge.

Hanfling, Oswald. 2002. Questions and Answers in Wittgenstein's Philosophy. In *Wittgenstein and the Human Form of Life*, 1–22. London: Routledge.

Johnston, Paul. 1989. *Wittgenstein and Moral Philosophy*. London: Routledge.

———. 1999. *The Contradictions of Modern Moral Philosophy. Ethics After Wittgenstein*. London: Routledge.

Kirchin, Simon. 2012. *Metaethics*. Basingstoke: Palgrave Macmillan.

Leich, Christopher, and Steven Holtzman. 1981. Introductory Essay. Communal Agreement and Objectivity. In *Wittgenstein. To Follow a Rule*, ed. Steven Holtzman and Christopher Leich, 1–27. London: Routledge.

Lovibond, Sabina. 1983. *Realism and Imagination in Ethics*. Oxford: Blackwell.

———. 2002. *Ethical Formation*. Cambridge, MA: Harvard University Press.

Pears, David. 1971. *Wittgenstein*. London: Fontana.

Rhees, Rush. 1965. Some Developments in Wittgenstein's View of Ethics. *The Philosophical Review* 74: 17–26.

Ripstein, Arthur. 1993. Questionable Objectivity. *Noûs* 27: 355–372.

Teichmann, Roger. 2001. The Functionalist's Inner State. In *Wittgenstein and Contemporary Philosophy of Mind*, ed. Severin Schroeder, 24–35. Basingstoke: Palgrave Macmillan.

Wiggins, David. 1998. Truth, Invention, and the Meaning of Life. In *Needs, Values, Truth*, 3rd ed., 87–137. Oxford: Clarendon.

———. 2006a. Objectivity in Ethics. In *Ethics. Twelve Lectures on the Philosophy of Morality*, 325–356. London: Penguin.

———. 2006b. Miscellanea Metaethica. In *Ethics. Twelve Lectures on the Philosophy of Morality*, 357–381. London: Penguin.

Wittgenstein, Ludwig. 1969. *The Blue and Brown Books. Preliminary Studies for the 'Philosophical Investigations*. 2nd ed. Oxford: Blackwell.

———. 1974. *On Certainty*. Oxford: Blackwell.

———. 1976. In *Wittgenstein's Lectures on the Foundations of Mathematics*, ed. Cora Diamond. University of Chicago Press: Chicago.

———. 1977. *Remarks on Colour*. Oxford: Blackwell.

———. 1978. *Remarks on the Foundations of Mathematics*. 3rd ed. Oxford: Blackwell.

———. 2009a. *Philosophical Investigations*. 4th ed., Trans. G.E.M. Anscombe, P.M.S Hacker, and Joachim Schulte. Oxford: Wiley-Blackwell.

———. 2009b. Philosophy of Psychology – A Fragment. In *Philosophical Investigations,* 4th ed., 182–244. Trans. G.E.M. Anscombe, P.M.S Hacker, and Joachim Schulte. Oxford: Wiley-Blackwell.

———. 2014. A Lecture on Ethics. In *Lecture on Ethics*, ed. Edoardo Zamuner, Ermelinda Valentina Di Lascio, and D.K. Levy, 42–51. Oxford: Wiley-Blackwell.

Wright, Crispin. 1980. *Wittgenstein on the Foundations of Mathematics*. London: Duckworth.

Part III
Wittgenstein's Later Philosophy and Contemporary Moral Philosophy

Chapter 6
Seeing Colour, Seeing Emotion, Seeing Moral Value

Defenders of moral perception have famously argued that seeing value is relevantly similar to seeing colour. Some critics think, however, that the analogy between colour-seeing and value-seeing breaks down in several crucial respects. Defenders of moral perception, these critics say, have not succeeded in providing examples of non-moral perception that are relevantly analogous to cases of moral perception. Therefore, it can be doubted whether there is such a thing as moral perception at all. I argue that, although the analogy between colour perception and moral perception does indeed break down in several crucial respects, that conclusion does not weaken the case of defenders of moral perception, because better analogies are available. If defenders of moral perception seek to draw support from an analogy, then seeing emotion will protect them better against criticisms than will seeing colour.

6.1 Introduction

It is often thought that we can perceive moral value at least in some cases, that we can see, for example, the needfulness of someone's situation, the goodness in a person, the rudeness of a gesture and the injustice of a deed.[1] Although nobody denies that we *say* that we perceive these things, some have argued that moral

[1] Sean McKeever and Michael Ridge rightly note that 'the cases in which we apparently can "just see" the moral status of certain actions tend to be ones in which the moral verdict is not particularly controversial. The perceptual model is less plausible for cases involving more controversy and complexity. I may be able to "just see" that killing a small child for fun is wrong but unable to "just see" whether abortion near the end of the second trimester would be wrong in a case of pregnancy due to rape where the child would suffer from a serious form of mental retardation and where the father opposes aborting the child' (McKeever and Ridge 2006: 78).

De Mesel Benjamin. 2016. Seeing Color, Seeing Emotion, Seeing Moral Value. *The Journal of Value Inquiry* 50: 539–555. Published by Springer. See https://link.springer.com/article/10.1007/s10790-015-9535-4.

B. De Mesel, *The Later Wittgenstein and Moral Philosophy*, Nordic Wittgenstein Studies 4, https://doi.org/10.1007/978-3-319-97619-8_6

perception is not really a form of perception, because there are too many significant differences between moral perception (say, seeing wrongness) and other, uncontroversial kinds of perception (say, seeing a certain shape). Defenders of moral perception are asked to provide examples of non-moral perception that are relevantly analogous to cases of moral perception in order to justify talk of perception in the moral case.[2]

David Wiggins (Wiggins 1998a, b) and John McDowell (1998b) have famously argued that seeing value is relevantly similar to seeing colour.[3] The analogy between colour-seeing and moral value-seeing is frequently invoked by those who seek to defend the possibility of moral perception. Some critics think, however, that the analogy between colour-seeing and value-seeing breaks down in several crucial respects (Blackburn 1985; Wright 1988). Defenders of moral perception, these critics say, have still not succeeded in providing examples of non-moral perception that are relevantly analogous to cases of moral perception. In short, the very idea of moral perception has been criticized by criticizing the analogy between colour perception and moral perception. If that analogy breaks down, then moral perception is thought to be in danger.

I will argue that, although the analogy between colour perception and moral perception may indeed break down in several crucial respects, that conclusion does not weaken the case of defenders of moral perception, because *better analogies are available*. Good candidates for an analogue of moral perception can be found in Wittgenstein's discussions of aspect perception (see Wittgenstein 1980a; b, 2009b). These discussions are multifaceted and complex, and Wittgenstein offers many (sometimes greatly varied) examples of aspect perception.[4] I will focus on one

[2] Moral perception can come in a variety of forms. For an overview, see Wisnewski (2015: 134). Jeremy Wisnewski points at the 'difference between acknowledging that a situation raises a moral issue and seeing how one ought to act within a situation where such an issue is raised', and adds that 'some might regard the sort of moral perception picked out by the first use (perceiving moral relevance) as common and plausible, but regard the second sense of moral perception (perceiving what one ought to do) as problematic' (Wisnewski 2015: 137). Peter Goldie (2007) distinguishes between perception of thick evaluative facts (such as seeing what is the kind thing to do) and perception of thin evaluative facts (such as seeing what is the right thing to do), and argues that moral perception is more plausible in the first case than in the second. In this chapter, I try to provide an answer to critics of the colour analogy who ask for examples of non-moral perception that are relevantly analogous to cases of moral perception. Thus, it suffices to point out that at least *some* cases of seeing moral value (not necessarily all of them) are relevantly analogous to at least *some* cases of seeing emotion, and I have chosen my examples with this goal in mind. One could, of course, try to defend an analogy between cases of *all* the varieties of moral perception and cases of emotion perception, but that project is much larger than mine.

[3] Strictly speaking, McDowell compares the perception of values to the perception of secondary qualities. But, as Peter Railton remarks, 'color has been the natural stand-in for "secondary quality" in most philosophical discussions of the analogy' (Railton 1998: 143).

[4] Avner Baz rightly remarks that there is 'no reason to suppose that Wittgenstein ever came as far as to form an idea of how his remarks on aspects, or some selection of them, may fall together to form some sort of a unified whole' (Baz 2011: 697). For an overview of different kinds and cases of aspect perception, see Schroeder (2010: 353) and Hausen and ter Hark (2013: 88–92). The difference between seeing aspects and seeing as (if there is one) is a controversial topic that will not concern us here.

group of examples: seeing emotion. We can see joy, grief, fear, or sadness (Wittgenstein 2009b: §227; Wittgenstein 1980b: §170; Wittgenstein 1981: §225).[5] Cases of seeing emotion, I will argue, provide better analogues of cases of seeing moral value than do cases of seeing colour.

I will proceed as follows. First, I briefly explain the motivation behind the colour analogy. Why has colour been thought to be an illuminating analogue of moral value? What are the relevant similarities between seeing colour and seeing moral value? I show that, with respect to these similarities, seeing emotion does at least the same work as seeing colour. It captures the initial motivation behind the colour analogy equally well. Second, I focus on dissimilarities between seeing colour and seeing moral value. In which crucial respects does the colour analogy break down? I show that, in contrast to seeing colour, seeing emotion is in all these respects relevantly similar to seeing moral value. I conclude that cases of seeing emotion provide a better model for moral perception than do cases of seeing colour. If defenders of moral perception seek to draw support from an analogy, then seeing emotion will better protect them against criticisms than will seeing colour.

6.2 Seeing Colour, Seeing Emotion, Seeing Moral Value: Similarities

What is the motivation behind the colour analogy? How is seeing moral value similar to seeing colour? Three similarities are particularly important.

(1) According to McDowell, colours are 'not adequately conceivable except in terms of certain subjective states' (McDowell 1998b: 136), that is, they essentially involve subjective responses. Herein lies the difference between secondary qualities such as colour and primary qualities such as shape: the latter are thought *not* to essentially involve subjective responses. Just like seeing colour but unlike seeing shape, seeing value cannot be adequately conceived except in terms of certain subjective states. Both colours and values are essentially *dependent on human subjectivity*.[6]

[5] While the analogy between moral perception and emotion perception is not itself totally original (see Audi 2013: 41, 58), the Wittgenstein angle is (as far as I know).

[6] McDowell is a dispositionalist about colour. For an interesting overview of how different colour theories may lead to different views on moral perception, see D'Arms and Jacobson (2006). They show how McDowell can be criticized not by criticizing the analogy (which is my focus), but by accepting the analogy and adopting a different colour theory which, if colours and values are analogous, then leads to a different theory about moral perception. Elizabeth Tropman (2010) argues, for example, that intuitionists can accept the colour analogy, although they do not believe that values are essentially dependent on human subjectivity, because they can defend the view that colours are not essentially dependent on human subjectivity either.

(2) Although, according to Wiggins and McDowell, colours essentially involve subjective responses, they are also in an important sense *independent and objective*. We do not project colours onto the world, they are there. McDowell writes:

> An object's being such as to look red is independent of its actually looking red to anyone on any particular occasion; so, notwithstanding the conceptual connection between being red and being experienced as red, an experience of something as red can count as a case of being presented with a property that is there anyway – there independently of the experience itself. (McDowell 1998b: 134)

In contrast to the experience of being in pain, for example, someone may experience something to be red without it actually being red or experience something to be red while believing or knowing that it is not red. There is room for ignorance and error about colours but not about one's own pain. Similarly, there is something independent and objective about values. It is not because something *appears* to be good, that it *is* thereby good, and we can be mistaken and ignorant about values. Wiggins claims that 'there resides in the combined objectivity and anthropocentricity of colour a striking analogy to illuminate [...] the externality that human beings attribute to the properties by which they evaluate things, people, and actions' (Wiggins 1998a: 108).

(3) In contrast to judgment and interpretation, perception suggests *directness and immediacy*. When we perceive something red, we do not infer on the basis of evidence that it is red, rather we spontaneously and non-inferentially see that it is so. Similarly, we immediately see that what the children are doing with the cat is wrong; we do not infer it on the basis of evidence.[7] Just like we do not infer what colour something is by comparing the actual colour to a colour sample, in some situations we do not infer what we have to do by applying a moral rule to a situation.

Although I largely agree with McDowell and his supporters on these points, I will not here defend the claims that their colour theory is right, that values are indeed objective while essentially involving subjective responses, or that ordinary colour perception is non-inferential. However, I do claim that *if* the colour analogy succeeds in showing at least partly what moral perception is like, *then* these three similarities are at the heart of its success. In aiming to show that Wittgensteinian aspect perception, and more specifically seeing emotion, will make for a better analogy, or at least for a viable alternative, it first needs to be ensured that these crucial similarities will not be lost. Can seeing emotion do the same work that seeing colour does? That will depend on the plausibility of the following claims:

(1*) Seeing emotion essentially involves subjective responses.
(2*) Emotions are independent and objective in the relevant sense.
(3*) There are cases in which we see emotion directly, non-inferentially.

---

[7] The immediacy of moral perception is emphasized in Starkey (2006) and Wisnewski and Jacoby (2007).

I have not defended (1), (2), and (3) about seeing colour, and I will not defend (1*), (2*), and (3*) about seeing emotion. What is important for my purposes is that (1*), (2*), and (3*) are relevantly analogous to what McDowell, Wiggins and their supporters say about value and colour, that they are *defensible* from a McDowellian/Wigginsian point of view. That is, *if* one accepts what they say about colour and value, *then* one will be inclined to accept (1*), that seeing emotion essentially involves subjective responses (it would be strange to hold that seeing colour involves subjective responses while seeing emotion does not), and (3*), that there are cases in which we see emotion non-inferentially (it would be strange to hold that we sometimes see non-inferentially that something is wrong, but that we never see non-inferentially that someone is sad). (2*) may seem more controversial. Are emotions not paradigm examples of the subjective, like pains? Is it not the case that I am sad whenever I feel sad, that I cannot feel sad while believing or knowing that I am not, that I cannot be mistaken or ignorant about my being sad? We can grant all that, because the analogy defended here is not an analogy between colours, emotions and values, but an analogy between *seeing* colours, *seeing* emotions and *seeing* values. I cannot see my own pains or emotions, I *have* them. It makes no sense to talk about similarities between seeing colours, seeing values and seeing my own emotions, because there is no such thing as seeing my own emotions. Thus, the considerations above are irrelevant to the analogy. Independence and objectivity in the relevant sense here mean that, when we see another person's sadness or pain, we do not project pain or sadness onto the person or produce it in her. We see what is there to see in persons, actions, or situations. To paraphrase McDowell: a person's being sad is independent of her or him actually looking sad to anyone on any particular occasion. Someone may see another person as sad without her actually being sad or see her as sad while believing or knowing that she is not sad. There is room for ignorance and error about sadness. It is not because a person *appears* to be sad, or because we believe her to be sad, that she *is* thereby sad. One could say, paraphrasing Wiggins, that there resides in the combined objectivity and anthropocentricity of others' emotions a striking analogy to illuminate the externality that human beings attribute to evaluative properties.

So, if the colour analogy does any work with respect to seeing value, the emotion analogy can do the same work. Those who support the colour analogy can support the emotion analogy too. Whether they do may depend on their theories of colour and emotion. Given a suitable theory of emotion, the emotion analogy will capture the crucial elements of the colour analogy equally well. It is interesting, in this respect, to note that McDowell is not only a realist about colour and value, but also a realist about other minds: 'We should not jib at, or interpret away, the commonsense thought that […] one can literally perceive, in another person's facial expression or his behaviour, that he is in pain, and not just infer that he is in pain from what one perceives' (McDowell 1978: 129).

Now that I have made plausible that three crucial similarities between colours and values, on which the colour analogy hinges, can be captured by the emotion analogy as well, I will discuss what I take to be some important dissimilarities between seeing colour and seeing value and show that, in these respects, seeing emotion is closer to seeing moral value than is seeing colour. Two points have to be noted in advance.

First, it is conspicuous that Wittgenstein himself repeatedly and explicitly refers to colour-seeing in his discussions of aspect-seeing, mostly attempting to point out that there are important differences between them and that the latter cannot be understood on the model of the former (Wittgenstein 2009b: §247). Most of the dissimilarities I will discuss can be found in Wittgenstein's work.

Second, most of the dissimilarities I will discuss have not gone unnoticed in discussions about moral perception. Simon Blackburn, for instance, goes so far as to call the colour analogy an 'evidently lame analogy' (Blackburn 1985: 17). Not only critics of moral perception, but also its defenders have pointed out dissimilarities. That there are dissimilarities does not as such harm their account. After all, they put forward an analogy and not an identity statement. It could be remarked that, if defenders of moral perception only want to bring out the three similarities I have mentioned, then criticisms directed at other points do not discredit the analogy, so critics trying to undermine the analogy by pointing at dissimilarities between seeing colour and seeing value have misunderstood what its defenders are saying. They put too much weight on the analogy or put the weight where it should not be put.[8] One could respond that some analogies easily break down when put under some weight and some do not, and that, although the dissimilarities may not have been denied by defenders of moral perception, there is a point in trying to prevent misunderstandings as well as unsound criticism. I will not pronounce on these matters here. My discussion of dissimilarities between seeing colour and seeing value is not meant to undermine the colour analogy. The dissimilarities are relevant in order to show that better analogies to seeing value are available. A better analogy, in this context, is one that (1) can do the same work as the original and (2) is similar to the object of comparison (in this case, seeing values) in important respects where the original analogue is dissimilar.

6.3 The Active Element

A frequent complaint about the colour analogy is that it provides a very passive model of moral perception. The analogy 'suggests a model of [...] evaluative experience as passive receptivity to the impingement of values' (D'Arms and Jacobson 2006: 212). But this is not how defenders of moral perception want to think of moral

[8] Simon Blackburn sees the colour analogy as 'the nub of the matter' (Blackburn 1985: 17) and Crispin Wright claims that moral realists rely on the comparison (Wright 1988: 1). According to Elizabeth Tropman, sensibility theory is 'motivated, in large part', by the analogy, and she remarks that 'the comparison fails alone to do all the work that McDowell sets out for it' (Tropman 2010: 31). Robert Audi remarks that 'some philosophers have developed an *extensive* analogy between moral properties and secondary qualities or indeed treated the former as a *subcase* of the latter' (Audi 2013: 49, my italics).

perception. Simon Kirchin emphasizes that colour sensation is a type of causal or mechanical process in which there is no room for freedom and control; 'colour responses just happen to us'. However, the same cannot be said in the case of ethics which requires an account that 'allows for the fact that human beings often develop their (immediate) responses consciously' (Kirchin 2012: 130–131). Andrew Fisher and Simon Kirchin formulate the problem as follows:

> There is some difference between value responses and colour responses. No matter what I think, I cannot help but see a red patch (in certain lighting conditions) as having a certain colour. I cannot decide to change what I think and, in the future, consciously try to respond differently. Value responses are different. Even if we initially respond to an action as being cruel, we can reflect on that response afterwards and try to justify it to ourselves and others as cruel. If no good justification is forthcoming, then we can change our judgement about that particular action and, over time, often change our natural, initial reaction to similar actions in the future. (Fisher and Kirchin 2006: 220)

In short, the difference between colours and values is that colours determine or fix our responses in a way that values do not.

Suppose that an object is placed before me and I am asked what colour it is. In normal circumstances, I can be mistaken about the colour, but I cannot be *wholly* mistaken. Confusion between red and pink is possible, but not between white and black. I cannot try (or decide to try) to change my perception of white into a perception of black, and there are no good reasons for trying to do so. The imperative 'Now try to see it as black' makes no sense. I cannot fail or succeed in seeing something white as black. Compare all this with the perception of moral value. I can be mistaken about the value I purport to see, and it would not be abnormal if I were wholly mistaken, if, for example, a good act appeared bad to me. I can try (or decide to try) to change my perception of moral value. Someone may urge me to try to see something as good which I had previously seen as bad, and I may succeed or fail.

The elements of freedom, control, conscious development, creativity, decision, and change can be grouped under what I will call the active element of value perception. While colour perception is dissimilar to value perception in this respect, aspect perception is not.[9] Wittgenstein stresses that aspect perception is subject to the will (Wittgenstein 2009b: §256). To say that it is subject to the will is not to say that we decide, in each and every case, to see an aspect or not to see it (as if we could not be struck by an aspect), but that it *makes sense* to order someone to try to see an aspect. Wittgenstein compares seeing an aspect to imagining in this respect: it makes sense to ask someone to try to imagine a tree, so imagining is subject to the will, but still the image of a tree can occur automatically. We can try to form an image of a tree and fail to do so, we can have images of trees and fail to get rid of them (see Hausen and ter Hark 2013).

---

[9] This is not to say that colour perception is entirely passive, that there is no active element in it at all. After all, we have seen that it crucially depends on subjective responses, and responses are not just reactions. P.M.S. Hacker lists some passive and some active elements of perception (Hacker 2013: 296–297).

If indeed, as Wittgenstein claims, aspect perception is subject to the will, it seems well-placed to mirror the active element of moral perception. Let us do the test with our leading example, seeing emotion. I can be mistaken about the emotion I purport to see, and it would not be abnormal if a happy face would appear sad to me (if, for example, I would see tears of happiness as tears of sadness). I can try (or decide to try) to change my perception of emotions. Someone may urge me to try to see the happy face covered with tears as a happy face, while I had previously seen it as a sad one, and in doing so I may succeed or fail.[10]

I conclude that aspect perception is active enough to capture the active elements of moral perception. At the same time, it is not too active. The passive element of colour perception and moral perception, shared by aspect perception, lies in the fact that I cannot choose or decide what I perceive. I cannot choose or decide to see a white table where there is a black one, to see a good act where there is a bad one or to see a happy face where there is a sad one. If the face is happy and I say that I see a sad face, I have made a mistake.

6.4 Education and Concept-Mastery

The active element in moral perception is closely linked to another aspect of it which is often emphasized by defenders of moral perception: adequate moral perception requires moral education, training and upbringing (McDowell 1998a, b; Wisnewski 2015). This Aristotelian idea is connected to what I said about trying to develop one's moral and emotional perceptions. One can try to do so by developing one's moral and emotional sensibilities. Of course, one can decide to develop one's colour sensibility too, and it is not unlikely that training and education will help one to discriminate colours better, to see nuances where other people do not, and so on. The development of moral and emotional perception on the one hand and colour perception on the other may be similar in many respects.

The similarities notwithstanding, there are important differences between colour education on the one hand and emotional and moral education on the other. In contrast to colour sensibilities, we *expect everyone* to develop their moral and emotional sensibilities and we accept that this development takes time and is never finished. It is true that we expect most people to be able to discriminate colours, but this process goes much faster, and there is no need or expectation that people will keep working on their colour sensibilities throughout their lives. People who have developed their moral and emotional sensibilities to an exceptionally high degree are said not only to see more, but also to understand more, to be wiser and more

[10] On the importance of the active element in seeing emotion, see Stout (2010: 39–40, 42). He remarks that, in order to see emotion, 'it helps if you are not too passive' (Stout 2010: 39).

mature than others. These terms are not used for persons with well-developed colour sensibilities.[11]

Wittgenstein emphasizes the role of education and upbringing in aspect-seeing (Wittgenstein 2009b: §168, §216). He further characterizes aspect perception as 'half visual experience, half thought', 'both seeing and thinking', or 'a fusion of the two' and 'the echo of a thought in sight' (Wittgenstein 2009b: §140, §144, §235). In order to be able to see certain aspects, such as emotions in a face, one needs to have mastered certain concepts, to have reached a certain level of intellectual sophistication (Baker 2004: 281; Glock 1996: 37).[12] According to Severin Schroeder, aspect-seeing is 'particularly concept-laden, typically more so than seeing shapes and colours' (Schroeder 2010: 360). Similar points are often made in discussions of moral perception, both by defenders and critics. Robert Audi notes that 'moral perception is possible for virtually every normal person with an elementary mastery of moral concepts' (Audi 2013: 121). Charles Starkey calls moral perception 'cognitively "thick" perception' and contrasts it to 'the "thin" characterization of perception as uncategorized seeing, hearing, smelling and so on' (Starkey 2006: 76). Crispin Wright explicitly refers to Wittgenstein's duck-rabbit. He claims that aspect perception, like moral perception, is possible only to a subject who has certain conceptual resources. This, according to Wright, makes colour perception a bad model, because such perception is 'up to a point at least, *raw*'. Wright concludes: 'So the suggestion is that there is no basis for describing an affective response as moral unless the subject gives evidence of the conceptual resources which would suffice to explain it as such' (Wright 1988: 12–13).

Michael Watkins and Kelly Dean Jolley describe moral perception as 'an intellectualized perceptual ability'. They add:

> We can say that these acquired skills [of moral perception], when they rely heavily on perception or are purely perceptual, are perceptual skills augmented by intellect. But to say this is not to say that the intellect adds something to what is seen, or somehow reshapes what is seen. Instead, it is to say that exercising the skill reveals something that is not revealed by unskilled, unfit, perceptions. Someone who exercises one of the skills correctly sees what a person without the skill does not see – but what is, nonetheless, there to be seen. Acquired perceptual skills provide information that unskilled perception cannot provide; but not because the skill adds something to what is seen. Correct exercises of the skill are revelatory, not creative. The mechanic who can tell what is wrong with a car by listening to it as it runs can hear something the non-mechanic does not hear. However, the mechanic's acquired perceptual skill does not create the mechanical trouble. (Watkins and Jolley 2002: 77)

[11] For reasons of space, not all the differences between colour education on the one hand and moral and emotional education on the other can be commented upon in this chapter. Goldie provides an account of certain differences between learning a virtue and learning a skill, and this seems like an interesting way to capture a difference between colour education and moral/emotional education. Learning to see moral value and emotion are, arguably, forms of (or close to) learning a virtue, while learning to see colour is a form of (or close to) learning a skill. One of Goldie's points is that (fictional) narratives have a more explicit and prominent role in learning a virtue than they have in learning a skill (Goldie 2007: 351–356).

[12] As I said, Wittgenstein provides many different examples of aspect perception. Not *all* aspect perception requires concepts or sophistication. See Baker (2004: 292, endnote 2).

Two things are remarkable here. First, what Watkins and Jolley say about moral perception, namely that the acquired skills of moral perception are perceptual skills augmented by intellect, but that this is not to say that the intellect *adds* something to what is seen, is almost exactly echoed by what Wittgenstein says about aspect perception: 'Is being struck [by an aspect] looking + thinking? No. Many of our concepts *cross* here' (Wittgenstein 2009b: §245). Thinking is not just *added* to seeing, but in aspect perception seeing and thinking are *inextricably interwoven.* Second, the fact that moral perception requires thought and concept-mastery does not make the term 'perception' any less appropriate. We see what is there to be seen, and we do not create the object of sight in thinking or imagination. Moral perception is 'revelatory, not creative'.

I conclude that, with respect to the need for education and concept-mastery, seeing colour and seeing value are in many respects different. In these respects, seeing emotion is closer to seeing value than is seeing colour.

6.5 Blindness

Seeing is conceptually connected to blindness: those who cannot see are blind. Are moral blindness and colour blindness relevantly analogous? And what about aspect blindness and emotion blindness?

If one is colour blind, one cannot see or discriminate between certain colours. Total colour blindness exists, but is rare. In most cases, what is wrong with the colour blind person is that her eyes are not functioning properly. The colour blind person cannot, for example, discriminate between red and green or cannot see red. In this sense, colour blindness is specific, that is, tied to certain specific colours. In another sense, colour blindness is general. *Whatever* the object of sight is, if it is red, one cannot see that it is red. Although it is possible that, in certain circumstances, one will be able to see that something is red, these circumstances will often be specifiable in general terms before the seeing occurs: when the lighting is such-and-such, when the object is made of such-and-such materials, and so on, you will be able to see red. A colour-blind person knows that she cannot see certain colours, and when somebody says 'But can't you see that this is red?', a typical answer will be 'No, I can't. I'm colour-blind. I cannot see red.'

Does total moral blindness exist? We can admit that it does, for example in psychopaths, and that it is, just like total colour blindness, rare. But a morally blind person is not someone whose eyes do not function properly. Although one cannot see the wrongness of an act if one's eyes are not functioning, this condition may not prevent one from hearing the wrongness, that is, from perceiving value in other ways. Perceiving moral value is not necessarily *seeing* moral value. According to Blackburn, the problem of the morally blind person is more aptly described as a defect of *character*. Moreover,

> [...] if our secondary-property-detecting mechanisms fail we know that immediately: it presents itself as a loss of immediately felt phenomenal quality, just as it does when the light fails or we stick cotton wool in our ears. There is no such loss when we become, say, corrupt. We cannot become corrupt overnight, and usually we cannot tell when we have done so. Indeed, it would be a hallmark of many kinds of moral blindness that this is so. The really coarse man thinks that he is perfectly in order, but that other people are too fastidious (recognizing that you have become really coarse is in this way self-refuting: the realization itself shows some residual delicacy). (Blackburn 1985: 14)

What is general and specific in moral blindness is different from what is general and specific in colour blindness. First, moral blindness is often tied to particular situations. Although I can see that many things are wrong, I cannot see that *this* is wrong. The blindness occurs in a particular situation, with a specific object of sight. Using a distinction made by Nigel Pleasants, one could say that the morally blind person is often someone who is *unable* to see something, but not in any general way *disabled*, while the colour blind person is unable to see something *because* he has a certain disability. While a disability prevents people from even trying to do what they are unable to do, an inability does not (Pleasants 2008: 110, 113). So while it makes no sense to urge the colour blind person to try to see the red, it does make sense, as we have seen, to urge the morally blind person to try to see things differently. The impossibility to see, the 'cannot', is of a different kind.

Although the situations in which a certain person will tend to show signs of moral blindness can sometimes be described in general terms ('She has a moral blind spot when it comes to the treatment of animals', or 'He is a narcissist, insensitive to the demands of others'), the information we need to be able to do so is not information about the lighting conditions and the functioning of his or her eyes. A morally blind person usually does not know that she cannot see certain moral properties, she will just deny that they are there to be seen. When somebody says 'But can't you see that this is good?', a typical answer by a morally blind person will be 'But it *isn't* good', without any reference to any general condition. This leads us, second, to what is general in moral blindness. For the most part, moral blindness is not tied to specific moral properties or values (this person cannot see goodness, this person cannot discriminate justice from injustice, etc.), although some persons can be said to see courage, for example, in general better than they can see injustice. Sometimes we cannot see that something is good; at other times we cannot see that something is courageous or unjust, and so on. For the most part, there are no specific moral properties that we cannot, in general, see.

Could there be a totally aspect-blind person? It is not entirely clear whether Wittgenstein thought there could be, but what is clear is that either there cannot be or that total aspect blindness is rare (Schroeder 2010: 366). Like the morally blind person, the aspect blind person is not someone whose eyes are not functioning properly. Although one cannot see sadness if one's eyes are not functioning, this condition will not prevent one from hearing the sadness, that is, from perceiving the emotion in other ways. Perceiving emotion is not necessarily *seeing* emotion, while perceiving colour is necessarily *seeing* colour. Wittgenstein remarks: 'Think of this too: I can only see, not hear, red and green – but to the extent to which I can see sadness, I can also hear it' (Wittgenstein 2009b: §220).

There is no loss of immediately felt phenomenal quality when we fail to see that a face is sad. We do not lose the ability to see emotion overnight, and usually we cannot tell when we have done so. What is general and specific in aspect blindness resembles moral blindness. Aspect blindness is usually tied to particular situations (Baker 2004: 281). Although I can see the sadness in many persons, I cannot see that *this* person is sad. The blindness occurs in a particular situation, with a specific object of sight. It makes sense to ask someone to try to see aspects differently and to criticize her for failing to do so. Although the situations in which a certain person will tend to show signs of aspect blindness can sometimes be described in general terms ('He has a blind spot when it comes to seeing grief'), the information we need to be able to do so is not information about the lighting conditions and the functioning of his or her eyes. An aspect blind person usually does not know that she cannot see certain aspects, she will just deny that they are there to be seen. When somebody says 'But can't you see that this face is sad?', a typical answer by an emotionally blind person will be 'But it *isn't* sad', without any reference to any general condition. This leads us to what is general in emotion blindness. For the most part, aspect blindness is not tied to specific aspects: this person cannot see sadness in a face, this person cannot discriminate joy from boredom, and so on, although some persons can be said to see grief, for example, in general better than boredom. Sometimes we cannot see that a face is sad, at other times we cannot see grief, boredom, and so on. For the most part, there are no specific aspects that we cannot, in general, see.

6.6 Appropriate Perceivers and Normal Observation Conditions

It is often thought that for something to be red is for it to appear red to appropriate perceivers under normal observation conditions, or that something is red if and only if it appears red to appropriate perceivers under normal conditions. The analogy of seeing colour to seeing moral value would then suggest that for something to be good is for it to appear good to appropriate perceivers under normal observation conditions, or that something is good if and only if it appears good to appropriate perceivers under normal conditions. There are several problems with this proposal. The main question is: what is meant by 'appropriate perceivers' and 'normal observation conditions'?

Critics of the analogy between seeing colour and seeing moral value contend that we can give a statistical interpretation of appropriate perceivers in the case of colour perception. The appropriate perceiver is the normal perceiver, a person with normal perceptual function, and this can be spelled out, according to Wright, as 'perceptual function of a kind which is actually typical of human beings' (Wright 1988: 15–16). What appears red to a supermajority of people under normal observation conditions, *is* red. If we compare the colour case to the moral case, they turn out to be different. We do not want the appropriate perceiver in the moral case to be the statistically typical or average person, because we want to leave room for the thought that a majority of people or the average person can be mistaken about certain moral matters (meat-eating, for example). According to Blackburn,

> [...] if we were to change so that everything in the world which has appeared blue came to appear red to us, this is what it is for the world to cease to contain blue things, and come to contain only red things. The analogue with moral qualities fails dramatically: if everyone comes to think of it as permissible to maltreat animals, this does nothing at all to make it permissible: it just means that everybody has deteriorated. (Blackburn 1985: 14)

What appears good to a majority of people under normal observation conditions, is sometimes *not* good. Remember, for example, how people used to think about slavery. So the statistical interpretation is not an option. While the appropriate colour perceiver is the normal, typical or usual perceiver, the notions of the typical or usual perceiver on the one hand and the one who appropriately perceives on the other hand seem to come apart in the moral case. If we do not want the appropriate moral perceiver to be the statistically average person, the only possible specification left seems to be that the appropriate moral perceiver is the one who sees things as they *ought* to be seen. What is good is what appears good to those who see things as they ought to be seen.

It should be remarked here that philosophers do not agree about the purported disanalogy between appropriate colour perceivers and appropriate perceivers of moral value. While critics of the analogy between seeing colour and seeing moral value, such as Wright and Blackburn, see a disanalogy on this point, others do not. Sydney Shoemaker (1994), for instance, argues that changing human physiology in such a way as to make blue things look red to (then) normal humans would *not* change their colour.[13] Maybe the appropriate colour perceiver is not the statistically average perceiver, or maybe there is no such thing as a typical or normal colour perceiver.[14] Whatever side one takes, I believe that it remains an open question whether, on this point, seeing emotion makes for a better analogue of moral perception than seeing colour does. Moreover, it is not clear to me what we should say about the appropriate emotion perceiver. Would it be possible for the majority of people to be mistaken and see a sad face as a happy one, and do we want to leave room for that possibility? In short, the question about appropriate perceivers does not have to be a problem for the colour analogy, and if it is, it is not obvious that seeing emotion does better or worse. Things seem different, however, with respect to normal or ideal observation conditions.

According to Wright, 'normal observation conditions' can be spelled out in the colour case, although a statistical interpretation is not plausible. Normal observation

[13] For a good discussion of the dispute, see Railton (1998).

[14] What do we make of the idea of an appropriate colour perceiver when confronted, for example, with the fact that women are better at discriminating among colours than men (see Abramov et al. (2012)), or with the fact that languages cut up the colour spectrum in different ways?

conditions are rather to be thought of as ideal or optimal observation conditions, and these can be specified:

> The conditions which actually usually prevail during winter in Spitzbergen, for instance, or in a normal photographic dark-room, are not suited for colour appraisal. A good description of conditions which are, optimally, so suited would be: conditions of illumination like those which actually typically obtain at noon on a cloudy summer's day out of doors and out of shadow. (Wright 1988: 16)

What Wright shows here is that, in order to see that something is red, certain conditions have to be met 'whose satisfaction does not directly depend on what the extension of colour predicates is' (Wright 1988: 24). That is, one does not have to know which things are red in order to know whether the conditions are met. The conditions do not refer to colour or to red.

Justin D'Arms and Daniel Jacobson remark that, while the challenge to identify observation conditions 'under which we are prepared to foreclose the possibility of ignorance and error [...] might be met satisfactorily in the case of colour, [...] it seems hopeless in the case of value' (D'Arms and Jacobson 2006: 201).

> Whatever standard conditions are chosen, we should not be inclined to grant that people under those conditions cannot be mistaken about values – unless the observers and circumstances are described simply as ideal, of course, in which case the characterization becomes trivial. (D'Arms and Jacobson 2006: 201–202)

While normal and ideal conditions for colour perception can be specified without referring to colour predicates, moral perception is 'a matter of meeting conditions the satisfaction of some of which is, irreducibly, a moral question' (Wright 1988: 24). That is, ideal conditions for moral perception are those in which an appropriate perceiver sees things as they *ought* to be seen. What is good is what appears good to those who see things as they ought to be seen in appropriate circumstances. However, this seems uninformative, and it shows why some have seen the threat of an infinite regress or vicious circularity in the account of moral perception given by McDowell and his supporters.[15]

What about aspect perception? With regard to normal observation conditions, there can be no doubt that these are very much like the normal observation conditions for moral perception. Normal observation conditions for seeing a face as sad, for example, seem to be those in which appropriate observers see the face as sad. We cannot but refer back to where we started. Those who think that the 'appropriate observers under normal conditions' account turns out to be circular in the case of moral perception (as opposed to colour perception), are likely to think the same when it comes to emotion perception. Thus, in this respect, moral perception and the perception of emotion are different from colour perception.

---

[15] For a brief explanation why the circularity need not be vicious, see Wiggins (1998b: 187–189) and Fisher and Kirchin (2006: 218).

6.7 Conclusion

The analogy between moral perception and colour perception breaks down in several crucial respects. In many of these, the perception of emotion is relevantly similar to moral perception. At the same time, seeing emotion keeps the similarities to seeing moral value that made seeing colour initially seem a useful analogue. Therefore, seeing emotion is on the whole a *better* analogue of seeing moral value than is seeing colour, although the latter has been a philosophers' favourite for decades.

What makes one analogy better than another is a disputed matter. Maybe the number of similarities does not matter all that much. What obviously (even trivially) matters, however, is the relevance of the similarities, and one could hardly deny that things like development and education are crucially relevant to moral perception. If, nevertheless, one does not accept the conclusion that the aspect analogy is better than the colour analogy, it should at the very least be recognized that it offers a good alternative to it. The fact that there are such alternatives may prevent us from, as Wittgenstein calls it, being held captive by a picture (Wittgenstein 2009a: §115), that is, by the idea that, if there is such a thing as moral perception, it has to be understood on the model of colour perception. If one holds this, one will think, as Blackburn and Wright do, that one can weaken the case of defenders of moral perception by attacking the colour analogy. The emotion analogy can show everything that the colour analogy was supposed to show, but is at the same time less vulnerable to these attacks. Therefore, the task of critics of moral perception may be more difficult than some of these critics have thought.

There are many things I have *not* done, and not tried to do, in this chapter. While I have been focusing on differences between colour perception on the one hand and moral and emotion perception on the other, I have not denied that there are interesting similarities between colour perception and moral perception or emotion perception, or between colours, emotions and moral values.[16] I have not tried to show (and how could I?) that seeing emotion is the best possible analogue of moral perception, nor that the emotion analogy never breaks down.[17] Wittgenstein insisted that

[16] There seems to be, for example, an interesting analogy between primary emotions and primary colours. Robert Plutchik, who developed the so-called 'wheel of emotions', suggesting eight primary emotions grouped on a positive or negative basis, writes that 'primary emotions can be conceptualized in a fashion analogous to a color wheel – placing similar emotions together and opposites 180 degrees apart, like complementary colors. Other emotions are mixtures of the primary emotions, just as some colors are primary and others made by mixing the primary colors' (Plutchik 2001: 349).

[17] It is important to note, however, that most of the difficulties for the aspect/emotion analogy are difficulties for the colour analogy too, so that they do not in any way harm the conclusion that the aspect analogy is a better analogy than the colour analogy. An example of such a difficulty is that moral perception is action-guiding in a way that aspect perception and colour perception are not (see Wright 1988: 8). While it is up to subjects whether they care about the colours or aspects they perceive, it seems impossible to perceive wrongness and not care about it (see Blackburn 1985: 15). This, however, is not necessarily a problem, as Wright recognizes, because it can be argued

philosophy should produce 'that kind of understanding which consists in "seeing connections". Hence the importance of finding and inventing *intermediate links*' (Wittgenstein 2009a: §122). I believe that seeing emotion is an intermediate link between seeing colour and seeing moral value, but there may be others. I have not discussed *all* the interesting similarities between seeing emotion and seeing moral value, and I suspect that the analogy is much richer than I have been able to bring out here.[18] I have not defended the claim that moral perception is genuine perception, nor have I supported any form of moral cognitivism or realism or objectivism or sentimentalism or their opposites. What I have shown is that (1) discussions about the perception of emotion, especially in the context of Wittgensteinian aspect perception, run strikingly parallel to discussions about moral perception and (2) that defenders of moral perception will be able to counter certain recurrent criticisms if they use seeing emotion instead of seeing colour as an analogue for seeing moral value.

It could be remarked here that seeing emotion may itself be a kind of moral perception.[19] If it is, then it cannot provide what critics of moral perception have asked for, namely, examples of non-moral perception that are relevantly analogous to moral perception. Jeremy Wisnewski (2015) claims that there are *moral* emotions, such as sympathy and love, and that perception of these emotions is therefore both moral perception and emotion perception. Thus, in some cases moral perception may be perception of emotion and *vice versa.* But the fact that moral perception and emotion perception may sometimes overlap does not mean (and Wisnewski recognizes this) that the two kinds of perception coincide, or that seeing emotion is a kind of seeing moral value or the other way round, and I am not aware of any convincing arguments to this effect. Not all moral perception is emotion perception (take, for example, seeing the injustice of a deed) and not all perception of emotion is moral

---

that moral perception *is* the perception of a cause for concern, that being the perception of a cause for concern is what is specific about moral perception (see McDowell 1998a; Starkey 2006: 86). Moreover, emotions seem much better placed as causes for concern than colours are. Another way to answer Wright and Blackburn on this point is suggested by Timothy Chappell, who argues that 'in evolutionary terms, what is hard to explain is not the representation that motivates, but the representation that does *not* motivate. As a matter of the history of our species, the (original) point of perceptual capacities in a tough world must usually have been to mandate response rather than to get hold of information for its own sake' (Chappell 2008: 434–435).

[18] (1) Wittgenstein contrasts colour disagreement with disagreement over the question of whether an expression of feeling is genuine or not (Wittgenstein 2009b: §§351–352). Seeing the genuineness of an expression of feeling is close to seeing emotion, but arguably already a form of moral perception (which is why I did not choose it as my leading example). So there may be interesting connections between disagreement about emotion and moral disagreement. (2) Wittgenstein asks whether there is such a thing as expert judgment about the genuineness of an expression of feeling (Wittgenstein 2009b: §355). The link between expert judgment about emotion and expert judgment about moral issues has been touched upon in this chapter, but can be worked out further. (3) Another similarity is that between what Wittgenstein calls the dawning of an aspect and Wiggins's repeated use of expressions such as 'lighting up' with respect to moral values. See Wiggins 1998a: 137 and 1998b: 207.

[19] I would like to thank an anonymous reviewer for this suggestion.

perception (take, for example, certain cases in which we see joy or sadness in someone's face). In this chapter, I have focused on cases of seeing emotion that I do not take to be cases of seeing moral value (and the other way round), cases of non-moral perception that are relevantly analogous to cases of moral perception. These cases of seeing emotion, I claim, do provide (or at least do better than cases of seeing colour in providing) what critics of moral perception have asked for.

At this point, one could remark that, however similar or dissimilar seeing emotion might be to seeing moral value, it does not show what the latter is like, because we hardly know what seeing emotion is like. The *explanans* does not explain. That could be true. But even if it does not explain, it helps, first, to emphasize or remind us of certain features of moral perception that are not captured by the colour analogy, so that we are less prone to be misled by that analogy. Second, it does something else that I find worth doing in philosophy. I agree with Wittgenstein when he says that 'Philosophy often solves a problem merely by saying: "*Here* is no more difficulty than *there*"' (Wittgenstein 1980a: §1000). He writes that 'the particular peace of mind that occurs when we can place other similar cases next to a case that we thought was unique, occurs again and again in our investigations' (Wittgenstein 2005: 307). According to Schroeder, the case 'then loses its disquieting uniqueness, its appearance of anomaly, and begins to look once more as common as it is' (Schroeder 2010: 364). Moral perception will all too easily appear unique and anomalous if one compares it to colour perception. Comparing it to emotion perception, by contrast, may make it look less disquieting and anomalous. And that, I presume, is something that defenders of moral perception will welcome.[20]

References

Abramov, Israel, et al. 2012. Sex and Vision II. Color Appearance of Monochromatic Lights. *Biology of Sex Differences* 3: 21.

Audi, Robert. 2013. *Moral Perception*. Princeton: Princeton University Press.

Baker, Gordon. 2004. The Grammar of Aspects and Aspects of Grammar. In *Wittgenstein's Method. Neglected Aspects*, 279–293. Oxford: Blackwell.

Baz, Avner. 2011. Aspect Perception and Philosophical Difficulty. In *The Oxford Handbook of Wittgenstein*, ed. Oksari Kuusela and Marie McGinn, 697–713. Oxford: Oxford University Press.

Blackburn, Simon. 1985. Errors and the Phenomenology of Value. In *Morality and Objectivity. A Tribute to J.L. Mackie*, ed. Ted Honderich, 1–22. London: Routledge & Kegan Paul.

Chappell, Timothy. 2008. Moral Perception. *Philosophy* 83: 421–437.

D'Arms, Justin, and Daniel Jacobson. 2006. Sensibility Theory and Projectivism. In *The Oxford Handbook of Ethical Theory*, ed. David Copp, 186–218. Oxford: Oxford University Press.

Fisher, Andrew, and Simon Kirchin, eds. 2006. *Arguing About Metaethics*. London: Routledge.

Glock, Hans-Johann. 1996. *A Wittgenstein Dictionary*. Oxford: Blackwell.

Goldie, Peter. 2007. Seeing What is the Kind Thing to Do. Perception and Emotion in Morality. *Dialectica* 61: 347–361.

[20] I am grateful to Nicole Hausen, Stefan Rummens and anonymous reviewers for comments on previous versions of this chapter.

Hacker, P.M.S. 2013. *The Intellectual Powers. A Study of Human Nature*. Oxford: Wiley-Blackwell.
Hausen, Nicole, and Michel ter Hark. 2013. Aspect Seeing in Wittgenstein and in Psychology. In *A Wittgensteinian Perspective on the Use of Conceptual Analysis in Psychology*, ed. Timothy P. Racine and Kathleen L. Slaney, 87–109. Basingstoke: Palgrave Macmillan.
Kirchin, Simon. 2012. *Metaethics*. Basingstoke: Palgrave Macmillan.
McDowell, John. 1978. On the Reality of the Past. In *Action and Interpretation*, ed. Christopher Hookway and Philip Pettit, 127–144. Cambridge: Cambridge University Press.
———. 1998a. Are Moral Requirements Hypothetical Imperatives? In *Mind, Value, and Reality*, 77–94. Cambridge, MA: Harvard University Press.
———. 1998b. Values and Secondary Qualities. In *Mind, Value, and Reality*, 131–150. Cambridge, MA: Harvard University Press.
McKeever, Sean, and Michael Ridge. 2006. *Principled Ethics. Generalism as a Regulative Ideal*. Oxford: Clarendon.
Pleasants, Nigel. 2008. Institutional Wrongdoing and Moral Perception. *Journal of Social Philosophy* 39: 96–115.
Plutchik, Robert. 2001. The Nature of Emotions. *American Scientist* 89: 344–350.
Railton, Peter. 1998. Red, Bitter, Good. In *Facts, Values and Norms. Essays Toward a Morality of Consequence*, 131–147. Cambridge: Cambridge University Press.
Schroeder, Severin. 2010. A Tale of Two Problems. Wittgenstein's Discussion of Aspect Perception. In *Mind, Method, and Morality. Essays in Honour of Anthony Kenny*, ed. John Cottingham and P.M.S. Hacker, 352–371. Oxford: Oxford University Press.
Shoemaker, Sydney. 1994. Self-Knowledge and 'Inner Sense. *Philosophy and Phenomenological Research* 54: 249–269.
Starkey, Charles. 2006. On the Category of Moral Perception. *Social Theory and Practice* 32: 75–96.
Stout, Rowland. 2010. Seeing the Anger in Someone's Face. *Proceedings of the Aristotelian Society. Supplementary Volume* 84: 29–43.
Tropman, Elizabeth. 2010. Intuitionism and the Secondary-Quality Analogy in Ethics. *The Journal of Value Inquiry* 44: 31–45.
Watkins, Michael, and Kelly Dean Jolley. 2002. Pollyanna Realism. Moral Perception and Moral Properties. *Australasian Journal of Philosophy* 80: 75–85.
Wiggins, David. 1998a. Truth, Invention, and the Meaning of Life. In *Needs, Values, Truth*, 3rd ed., 87–137. Oxford: Clarendon.
———. 1998b. A Sensible Subjectivism? In *Needs, Values, Truth*, 3rd ed., 185–214. Oxford: Clarendon.
Wisnewski, Jeremy J. 2015. The Case for Moral Perception. *Phenomenology and the Cognitive Sciences* 14: 129–148.
Wisnewski, Jeremy J., and Henry Jacoby. 2007. Failures of Sight. An Argument for Moral Perception. *American Philosophical Quarterly* 44: 229–244.
Wittgenstein, Ludwig. 1980a. *Remarks on the Philosophy of Psychology*. Vol. 1. Oxford: Blackwell.
———. 1980b. *Remarks on the Philosophy of Psychology*. Vol. 2. Oxford: Blackwell.
———. 1981. *Zettel*. 2nd ed. Oxford: Blackwell.
———. 2005. *The Big Typescript*. Oxford: Wiley-Blackwell.
———. 2009a. *Philosophical Investigations*. 4h ed., Trans. G.E.M. Anscombe, P.M.S Hacker, and Joachim Schulte. Oxford: Wiley-Blackwell.
———. 2009b. Philosophy of Psychology – A Fragment. In *Philosophical Investigations*, 182–244. 4th ed., Trans. G.E.M. Anscombe, P.M.S Hacker, and Joachim Schulte. Oxford: Wiley-Blackwell.
Wright, Crispin. 1988. Moral Values, Projection, and Secondary Qualities. *Proceedings of the Aristotelian Society. Supplementary Volume* 62: 1–26.

Chapter 7
Do Moral Questions Ask for Answers?

It is often assumed that moral questions ask for answers in the way other questions do. In this chapter, moral and non-moral versions of the question 'Should I do x or y?' are compared. While non-moral questions of that form typically ask for answers of the form 'You should do x/y', so-called 'narrow answers', moral questions often do not ask for such narrow answers. Rather, they ask for answers recognizing their delicacy, the need for a deeper understanding of the meaning of the alternatives and the fact that moral decisions are, as Gaita formulates it, 'non-accidentally and inescapably' the agent's to make. In short, moral questions often ask for a kind of answer that is highly different from the kind of answer non-moral questions ask for. In presupposing the ideal answer to a moral question to be a narrow answer, moral philosophers have tended to overlook this.

7.1 Introduction

Consider two situations:

(1) Alexander is a researcher. He is on his way back to his hotel, after a boring conference day in a city that he has never been to before. On the way, however, he gets lost. At a crossing, he comes to the conclusion that he must either go straight or turn left.
(2) Jonathan is Alexander's colleague. On the way to his hotel, after the same boring conference day in a city that he also has never been to before, Jonathan

---

De Mesel Benjamin. 2015. Do Moral Questions Ask for Answers? *Philosophia. Philosophical Quarterly of Israel* 43: 43–61. Published by Springer. See https://link.springer.com/article/10.1007/s11406-014-9565-3.

B. De Mesel, *The Later Wittgenstein and Moral Philosophy*, Nordic Wittgenstein Studies 4, https://doi.org/10.1007/978-3-319-97619-8_7

comes to the conclusion that he must either leave his wife or abandon his work in cancer research.[1]

Both Alexander and Jonathan face a problem about what they should do. Jonathan's, however, is a moral problem, while Alexander's is not. Suppose that both men come to the conclusion that they cannot solve their respective problems and that they need help. A normal thing for Alexander to do would be to approach a passer-by and ask: 'Excuse me, I am looking for the Chelsea Hotel. Should I go straight or turn left here?' If Alexander is lucky, the passer-by would answer something like: 'You have to turn left here. The hotel is on the right side of the street.' A normal response for Alexander would be to thank the passer-by, immediately take her advice and turn left. We will call this scenario S1.

Suppose now that Jonathan approaches a passer-by and asks: 'Excuse me, I am looking for a happy life. Should I leave my wife or abandon my cancer research?'[2] And suppose that the passer-by answers: 'You should definitely leave your wife.' Suppose that Jonathan thanks the passer-by, immediately takes her advice and leaves his wife. We will call this scenario S2. While S1 is a realistic scenario, S2 is obviously absurd. Jonathan's question, the passerby's answer and Jonathan's reaction are neither normal nor realistic. And surely, we would not say that Jonathan is lucky if the passer-by had answered in the way that she does in S2.

The question, then, is as follows: why do we think that Alexander's question, the passerby's answer and Alexander's reaction to it are normal, while we think that Jonathan's question, the passerby's answer and Jonathan's reaction are absurd? In Sects. 7.2 and 7.3 of this chapter, I will try to answer that question, drawing on insights from Ludwig Wittgenstein and Raimond Gaita, and explain where the crucial differences between S1 and S2 lie. In Sect. 7.4, I will argue that these differences indicate characteristics of moral questions that have been overlooked all too often. I will conclude by arguing that these characteristics suggest that moral questions often do not ask for answers in the way that other questions do.

7.2 Why Is S2 Absurd?

There are many different kinds of questions. For reasons of space and clarity, I will only be concerned here with questions of the following form: 'Should I do x or y?' As the examples of Alexander and Jonathan show, these questions can be non-moral as well as moral in nature. In a *broad sense* of 'answer', an answer is 'what is said, written or done as a reaction to a question' (Soanes and Stevenson 2006: 64).

[1] The example is Rush Rhees's. Rhees discussed it with Wittgenstein (Rhees 1965: 22).

[2] Instead of a happy life, Jonathan might say that he is looking for a virtuous life, a moral life or a good life. He might also say that he wants to act rationally or dutifully, that he wants to do whatever has the best consequences or that he wants to treat people as ends in themselves, depending on what he takes to be fundamental matters of moral concern. Whatever he says here, the scenario remains absurd.

According to this definition, non-moral as well as moral questions of the form 'Should I do x or y?', in so far that they ask for reactions, ask for answers.

Suppose that one of my children is very ill and I ask my wife: 'Should I visit Alice in the hospital or should I go to the cinema?' She could have any number of reactions. For instance, she could say, 'How dare you even ask such a stupid question!' She could say nothing at all and just stare at me indignantly for daring to ask such a stupid question, or she could even slap me in the face. These reactions qualify as answers in a broad sense, but they do not qualify as answers in a narrow sense. An answer in the *narrow sense* is one in which the only possible answers to a question of the form 'Should I do x or y?' are as follows: 'You should do x', 'You should do y', 'You should not do x', 'You should not do y', 'You should do both' (if x and y are not mutually exclusive), and 'You shouldn't do either' (if x and y do not exhaust the options). It is clear that this is the kind of answer (especially in the form 'You should do x' or 'You should do y') that Alexander is looking for. The absurdity of Jonathan's conversation leads one to doubt whether the same holds true for him.

This doubt, of course, does not of itself lead to the conclusion that the difference between S1 and S2 is due to the fact that Jonathan's question, *because* it is a moral question, does not ask for an answer in the narrow sense. Other differences between the situations may account for the absurd character of Jonathan's conversation. A difference suggesting itself is that it is highly likely that the passer-by in S1, whom we will call Ada, knows everything she has to know in order to give a satisfying answer to Alexander's question. She may very well be a local who knows the area, meaning that she may have all the factual knowledge that she needs to properly help Alexander. Therefore, it is reasonable for Alexander to immediately take her advice.

Things are different in S2. We will assume that Ada does not know Jonathan and has never seen him before. She clearly lacks any knowledge about Jonathan and of Jonathan's situation. Does he have any children and how old are they? What is his relationship like with his wife? Does he have any colleagues who could continue his work if he decided to abandon it? Etc. One might think that, the more information that Ada has about Jonathan, the less absurd S2 would be and, consequently, the more appropriate it would be for Jonathan to accept a 'narrow' answer. Imagine, then, the following variation of S2:

Jonathan: Excuse me, I am looking for a happy life.[3] Should I leave my wife or abandon my cancer research?

Ada: I'm very sorry, but I don't know you and your situation well enough to be able to answer your question.

Jonathan: I understand. Fortunately, I have been keeping a detailed diary throughout my life. I have kept all of my letters and a biography has even been written about me. There is a biography about my wife too, and her

[3]One could argue that this opening sentence makes every scenario highly absurd. Hence, it becomes difficult to show that S3 is any more or less absurd than S2. I would like to emphasize that even if the opening sentence of S2 and S3 were quite innocent (such as 'Excuse me, I would like to ask you something'), S3 would not be any less absurd than S2. The fact that Ada recognizes the need for more information does not as such affect the scenario's degree of absurdity.

diaries have been published recently. I will make sure that you get all of the information you might possibly need.

Ada: Very well. I will take some time to study everything and I will send you an e-mail when I'm ready.

A month later, Jonathan receives an e-mail from Ada, stating that she has studied everything and that she has come to the conclusion that he should definitely leave his wife. Let this scenario be S3. If Jonathan now were to leave his wife immediately, would that be less absurd than his reaction in S2?

No, one might say, because it would not be reasonable for Jonathan to trust Ada's judgment. Even if she proves in her e-mail that her knowledge of Jonathan's situation is impeccable, she may not know what to do with it, how to process the information. Even if we assume that she has all the knowledge that she needs about Jonathan, she may lack the proper moral expertise or moral knowledge, i.e. knowledge about how, starting from a heap of information, one can come to a reasonable moral judgment. If that accounted for the strangeness of the conversation above, we could dissolve this problem by simply stipulating that Ada is a distinguished professor of ethics with countless publications in *Ethics* and *The Philosophical Review*. Jonathan may have read and appreciated her most famous book and recognized her from her photograph on the back flap. When seeing her on the street, he decides to ask his question because he takes her not only to have ample knowledge of the history of ethics and of ethical theories, but also to know what is morally relevant in concrete, everyday cases, as attested to by the eloquent way in which she applies her theory to concrete, everyday examples in her masterly book. Now, even if it was the case that Ada was a professor of ethics, or an otherwise morally competent and wise person whose expertise and knowledge of the subject it would be most reasonable for Jonathan to trust, would it be reasonable for Jonathan, after receiving the e-mail, to immediately tell his wife that he is leaving her?

No, I would say, even if the answer is maximally reliable, exhibits an impressive knowledge of Jonathan's situation and shows a strong capacity for moral reasoning that Jonathan would have no reason for distrusting.[4] Of course, a distinguished professor of ethics would not talk to Jonathan like Ada does in S3. Why not though, if the only thing that she needs to be able to answer Jonathan's question is some information or knowledge, some input for her ethical theories to generate an output in the form of a reliable moral judgment? Apparently, what is at stake here is not the reliability of Ada's answer. If Jonathan is reasonable, he will not reject the answer or refrain from immediately acting upon it because the answer is unreliable or because he suspects it to be wrong. That is why making Ada a professor of ethics does not fundamentally change the situation.[5]

[4] This is known as the problem of moral deference. See Hopkins (2007), Jones (1999), McGrath (2009).

[5] Those who do not think of professors of ethics as paradigms of reliability when it comes to dealing with moral questions are free to change 'professor of ethics' by 'wise person'.

Apart from differences in *reliability of the passersby and of their answers*, apart from the possibility that they *lack factual or moral knowledge*, what could account for the absurd character of S2, while S1 seems perfectly normal? Another possible answer is this: Alexander's question is very easy, but Jonathan's is enormously difficult. Thus, it is this difference in *levels of difficulty or degrees of complexity* that explains the different ways in which we respond to S1 and S2. If that were true, changing Alexander's question into an enormously difficult non-moral question would make his conversation as unreal as Jonathan's in S2. Imagine, then, the following variation of S1.

Alexander: Excuse me, I am looking for ways to maximize my investments. Should I buy shares in company x or in company y?

Ada: You should buy shares in company x.

So far, this scenario is at least as strange as S2. The difficulty of the question makes it unreasonable for Alexander both to ask the question to a stranger and to trust her answer. Let us change the scenario, though, as we did with S2; that is, let us stipulate that Alexander recognizes Ada as a distinguished professor of economics. If, in this scenario, which we will call S4, Ada promises to take some time to study everything, and if Alexander, after a month, receives an e-mail containing the conclusion 'You should buy some shares in company x', would it be unreasonable for Alexander to immediately do so?

S4 suggests that it is neither the difficulty nor the complexity of the question that accounts for the absurdity of S2.[6] Another option is to say that it is not so much the difficulty, but the *importance* of the questions that makes all the difference. Indeed, 'Should I go straight or turn left here?' is probably far less important for Alexander than 'Should I leave my wife or abandon my research?' is for Jonathan. The consequences of a wrong answer are probably far easier for Alexander to deal with than they would be for Jonathan. This does not necessarily have to be the case, though. Suppose that Alexander has just received a phone call from his wife, telling him that one of his children is seriously ill, that he must come home as soon as possible and that he can take a plane in an hour. Alexander has to go to his hotel to take his passport. His question now becomes very important, and Ada, seeing how nervous he is, may well understand that. However, that obviously would not incline her not to give a straightforward answer or to think about it longer. Also, it would be strange if Alexander did not take her advice immediately and did not turn left.

I conclude that the difference between S1 and S2 cannot be adequately explained by referring to:

(1) The probable lack of factual and/or moral knowledge and/or reasoning capacities of the passer-by in S2 compared to S1 and, consequently, the probable

[6] If some would object to this example that the stock market question is not difficult enough in comparison to Jonathan's question, they are free to make Alexander's question more difficult. It will not fundamentally change the contrast with S2.

unreliability of her narrow answer; that is, the probability of its being false, the chance of there being a better narrow answer;
(2) The difficulty or complexity of the question in S2 compared to S1;
(3) The importance of the question in S2 compared to S1.

If one of these factors, or all three together, were enough to explain the difference between our reactions to S1 and S2, there would be no reason to doubt that Jonathan's question asks for a narrow answer of the form 'You should do x', because many difficult (2) and important (3) questions, even if they cannot be adequately or reliably answered without extra information or knowledge (1), *do* ask for narrow answers. It would be normal for Alexander to be disappointed or dissatisfied if Ada did not give a narrow answer to his difficult and important question that required so much information ('Should I buy some shares in company x or y?'). If someone said to him 'I cannot help you', Alexander would be justified in thinking that his question had not been answered at all (on a narrow conception of 'answer') or at least that this was not the answer (a broad answer) that he was looking for. I will argue that this is not necessarily the case for Jonathan.

7.3 Why S2 Is Absurd

7.3.1 Jonathan's Question

If not for the reasons mentioned above, why is S2 so absurd? Let us first focus on Jonathan's question: 'Should I leave my wife or abandon my research?' It seems as if, although the question would be meaningful in certain contexts, we cannot make sense of that question within the context of S2, as if language had gone 'on holiday' here (Wittgenstein 2009: §38; see also Wittgenstein 2009: §117). The question seems to lack the typical surroundings and circumstances, the environment in which it would actually be used and is 'at home' (Wittgenstein 2009: §116). What are these 'typical surroundings and circumstances', and do they indeed account for the absurdity of S2?

First, there is something inappropriate in Jonathan's asking this question to a total stranger. If Jonathan has to ask it at all, it seems that it would be better to ask it to one of his friends, to somebody who he knows and who knows him. But is that not exactly the problem we compensated for in S3? If Jonathan was to provide Ada with his full biography, providing the stranger with every bit of information that she might possibly want about Jonathan, would we not have to admit that she *knows* Jonathan? I do not think so. The crucial difference here is between *knowing everything about* Jonathan and *knowing* Jonathan or *knowing* Jonathan *personally,* between knowing *what* he is and knowing *who* he is. Hannah Arendt writes:

> The moment we want to say *who* somebody is, our very vocabulary leads us astray into saying *what* he is; we get entangled in a description of qualities he necessarily shares with others like him; we begin to describe a type or 'character' in the old meaning of the word, with the result that his specific uniqueness escapes us. (Arendt 1958: 181)

Here, Arendt captures the intuition that one can know everything about Jonathan but not know him, and that one can know Jonathan but not know everything about him.[7] The former happens when one knows about Jonathan only through 'vocabulary', through knowledge of paradoxically impersonal personal properties or through the biography and the diaries we have been referring to in S3. The stranger who has read every bit of information about Jonathan still lacks something essential to say that she properly knows him; what she lacks, is a *shared history* with him. They have had no life together. They have not gone through things together, they have not seen each other living their lives. They have not shared any experiences, indeed, they have never even seen each other before.

I do not want to suggest that Jonathan can only ask this question of friends, people that he likes or people that he knows very well. What is minimally required is *some* shared history. For instance, it may be reasonable to ask a stranger whom one considers morally competent or wise for moral advice. Take the people asking Father Zosima for advice in Dostoevsky's *The Brothers Karamazov.* It would not be reasonable for them to take Zosima's advice unless he made some effort to get to know them. Of course, that is precisely what Zosima does. He talks to people, gets to know them personally and, as it befits the wise, is very good at getting to know people personally in a short period of time. A wise or morally competent person is not someone who is able to produce a correct moral judgment on the basis of impersonal information and moral theories. The need for a shared history and personal knowledge in Jonathan's case explains why it would be ridiculous and of no help to Jonathan to answer his question as follows: 'I'm very sorry, but I do not feel competent to answer your question. Maybe you should post your question on a specialized internet forum. Or take a subscription to *Ethics.*' Would it be equally stupid to advise Alexander to take a subscription to a distinguished financial magazine or to post his question on a specialized internet forum in order to see in which company he should invest his shares? The need for a shared history and personal knowledge further explains why we would reject the idea of a machine answering Jonathan's question: could we even imagine trusting a 'moral machine' with all the information about our lives and circumstances as input and a moral judgment as output? Compare this with the role played by computers in stock market advice or with the idea of a machine showing Alexander the way to his hotel. These machines exist, of course, and we trust them and act upon their 'advice' all the time. A moral GPS, by contrast, upon the advice of which it would be reasonable to act immediately, does not even figure in science-fiction.

---

[7] See, in this respect, also Wittgenstein (1992: 89–90): 'It's important, for instance, that one must "*know*" someone in order to be able to judge what meaning is to be attributed to one of his expressions of feeling, and yet that one cannot describe what it is that one knows about him.'

7.3.2 Mary's Answer

Let us take the need for a shared history and personal knowledge into account and stipulate that Jonathan asks his question to someone he knows and shares some history with; say, to his friend Mary. Why would it be inappropriate for Mary to answer 'You should definitely leave your wife'? First, there is something wrong with giving such a quick answer to Jonathan's question. Things are different for Alexander, particularly in the situation where his child is seriously ill. As to his question of whether to buy shares in company x or y, of course, Alexander may not expect such a quick answer. If he asks this question to a world-leading specialist who happens to know the companies very well, though, and if she stresses the word 'definitely' in saying that Alexander should *definitely* buy shares in company y, would it be unreasonable for Alexander to trust the answer and immediately act upon it? Of course, the economist may not happen to know the companies very well, and a quick answer may not be forthcoming. The point is that there seems to be no equivalent notion of 'happening to know' in moral cases. What if Mary replied to Jonathan's question as follows: 'This is one moral question that I happen to know the answer to. You should definitely leave your wife.'?

The quickness of the answer seems to be a problem in Jonathan's case. If someone were to answer his question as quickly as Mary does, we would think that she had not seriously considered the question or thought it through. Indeed, this is a situation in which *seriously considering the question* is necessary.[8] This is not because the question is particularly difficult. Computers can solve immensely difficult questions in less than a second and we have no problem with the fact that they do not seriously consider things. Computers simply cannot be said to seriously consider questions. This is one of the reasons why we would never seriously ask a computer a question like Jonathan's and why we would never trust and act upon whatever answer that it provided to such a question. In order for Mary's answer to be acceptable to Jonathan, Mary's *attitude* is important. Jonathan has to believe that her attitude (something computers do not have) is marked not only by seriousness, but also by *engagement and commitment*, by a genuine *interest* in the question. Alexander, for sure, would not deplore the passerby's or the economist's having such an attitude, but he need not believe that they have it in order to trust and act upon their answers. On this point, I agree with Gaita:

> We would not seek moral advice from someone whom we knew to be morally jaded. Being scientifically jaded, however, in the sense of one's interest in science having 'gone dead' on one, is of itself no bar to a scientist's authority to speak in his field, provided only that his memory is good and that he has not been jaded for too long, for if he had been, we would be doubtful whether he had sufficient energy or interest to keep up with his subject. (Gaita 2004: 103)

The first problem with Mary's answer is that its quickness seems to preclude a certain attitude that Jonathan's question asks for. The second problem is that it is

[8] On the importance of seriousness in characterizations of the moral, see Gaita (2004: *passim*).

somehow too naked, too hard, too head-on, too straightforward, too direct and, indeed, too narrow. If Jonathan were asking for a narrow answer, then a maximally reliable narrow answer, given by someone who he knows and who knows him, someone who has all the knowledge and reasoning capacities that she needs, someone who is aware of the difficulty and importance of the question, someone who is interested and committed and has seriously considered the question, would be what he wanted. But even if Mary was someone like that and if she took her time to answer, Jonathan would not be satisfied with the naked answer. It would be perfectly reasonable for him to ask: 'How do you know that? Why would you say that?'. In other words, he would want to know Mary's *reasons* for saying what she says: does she think his research would save many lives, that nobody could take it over from him, that his wife will find another man, that he will regret it if he does not leave her, etc. Does this 'wanting to know the reasons' distinguish Jonathan's situation from Alexander's? It would be strange if Alexander asked Ada why she thinks he should turn left, but would it not be normal if he would want to know the economist's reasons for recommending that he buy shares in company x? If Alexander was curious or interested in economics, it certainly would. The point is, however, that Alexander does not have to be curious or interested in economics. Perhaps he just wants to know what he should do with his money. He does not need to be interested in the reasons; he could just as well say 'I am not interested in her reasons. What counts is that her answer is right.'

Things are certainly different for Jonathan here. The difference is quite similar to the one Wittgenstein points at in 'A Lecture on Ethics':

> Supposing that I could play tennis and one of you saw me playing and said 'Well you play pretty badly' and suppose I answered 'I know, I am playing badly but I do not want to play any better', all the other man could say would be 'Ah then that's all right'. But suppose I had told one of you a preposterous lie and he came up to me and said 'You are behaving like a beast' and then I were to say 'I know I behave badly, but then I do not want to behave any better'. Would he then say 'Ah, then that's all right'? Certainly not; he would say 'Well, you *ought* to want to behave better'. (Wittgenstein 2014: 44)

Similarly, if Alexander was to say 'I don't want to know her reasons', we could just say 'Ah, then that's all right'. We could not say the same for Jonathan, though, if he decided that he did not want to know Mary's reasons. Reasons play a different role in Alexander's and Jonathan's respective cases. Drawing from Gaita, I would say that Jonathan is asking his question not only because he wants to find out what to do, but also because he wants to *achieve a deepened understanding of the meaning of the alternatives*: 'A deepened understanding of the *meaning* of our actions […] is a mark […] of what we should understand by having arrived at a "right answer" to a moral question' (Gaita 2004: 264). If Mary says that Jonathan should leave his wife because 'You will save many lives', then she is urging Jonathan to think of his leaving his wife as of something that saves lives. To leave his wife would not just have the saving of lives as a consequence, it would *mean* or *be* to save lives. Mary portrays the alternatives in a certain light so that Jonathan will understand them in a certain way. The more Jonathan finds that this way of putting things deepens his understanding of the situation, the more likely he is to think that his question has been satisfactorily answered, even if no direct or straightforward answer is forthcoming. Imagine Mary saying things like 'You will save many lives and you will regret it if you do not leave her. On the other hand, her life will be ruined and she may never forgive you' without her giving a narrow answer. Compare this with her saying 'You should definitely leave your wife'. Jonathan may well prefer the first kind of answer.[9] That is, reasons or explanations and novel perspectives on the situation do not play a subordinate role in Jonathan's case as they do in Alexander's, where they serve as a kind of road leading to the final destination which is the narrow answer. The ultimate aim here is the narrow answer, and those who are curious may optionally ask 'How did you get there?'[10] Jonathan may not want Mary to bring him to a final destination. Instead, he wants her to show paths that he may take and that he may not have known or thought of.

In this context, we should discuss another possible answer to Jonathan's question. In a realistic scenario, Mary could answer 'Should I leave my wife or abandon my research?' by asking why Jonathan believes that he has to either leave his wife or to abandon his research. She would thereby invite him to reconsider the situation and the way in which he frames it.[11] Are the two options really mutually exclusive, as Jonathan seems to think? Mary's reaction to Jonathan's question consists in challenging the very terms in which the question is asked (and the absurdity of S2 can be explained partly by referring to the fact that the passer-by straightforwardly and without hesitation accepts the terms of the question). This challenge may lead Jonathan to achieve a deepened understanding of the meaning of the alternatives (perhaps they are not mutually exclusive at all, perhaps they are not really alternatives, perhaps there are other options, etc.). It counts as a broad answer, but not as a narrow one. In the end, of course, Mary may give a narrow answer to Jonathan's

---

[9] An anonymous reviewer remarks that 'whether Jonathan prefers "You will save many lives" over "You should definitely leave your wife" depends on whether or not he is someone who wants to make his own decisions and take responsibility for them, i.e. someone who values autonomy. Certainly, not everybody does.' I agree and, as I will repeat in Sect. 7.4, I readily admit that there are cases in which a person is asking for a narrow answer. Indeed, there are situations in which it would be absurd to withhold a quick, straightforward moral judgment of the form 'You should do x'. Far from claiming that the latter is always unwelcome in moral cases, my claim is that one cannot assume that it is always appropriate. It suffices for my purposes, then, to show that we can clearly imagine a situation in which and a person for whom such an answer would certainly not be appropriate. Therefore, we could simply stipulate that Jonathan values autonomy without harming the argument.

[10] The idea that reasons and explanations are not optional in moral cases as they are in many non-moral cases (that the hierarchy between reasons and conclusions is different in moral and non-moral cases) is present in Bernard Williams's famous remark that 'anyone who is tempted to take up the idea of there being a theoretical science of ethics should be discouraged by reflecting on what would be involved in taking seriously the idea that there were experts in it. It would imply, for instance, that a student who had not followed the professor's reasoning but had understood his moral conclusion might have some reason, on the strength of his professorial authority, to accept it [...] These Platonic implications are presumably not accepted by anyone' (Williams 1995: 205).

[11] I would like to thank an anonymous reviewer for this suggestion. Cora Diamond emphasizes that 'in moral life we are not in general confronted with cases the description of which can be taken to be simply given' (Diamond 2002: 238). See also Diamond (1991) and Gaita (2004: 64–73).

question. As we have seen, 'You should do both' and 'You should do neither' qualify as narrow answers.[12] It is not my aim to argue that narrow answers are *never* appropriate. I only want to make a case for the much weaker claim that they are *not always* appropriate, that it should not be assumed that moral questions ask for narrow answers. The point here is that, if Jonathan wants to achieve a deeper understanding of the meaning of the alternatives, a broad answer is often enough. A narrow answer is not necessarily what Jonathan asks for.

The above considerations regarding the primacy of reasons and understanding over narrow answers in Jonathan's case suggest that Mary's narrow answer 'You should definitely leave your wife' indicates her misunderstanding of the kind of question Jonathan has asked her and of the kind of answer he is looking for. We have called Mary's answer not only too narrow and too naked, but also too hard and too direct. Giving ample explanations and providing reasons may help to compensate for the narrowness and the nakedness of the answer. Nevertheless, if Mary were to conclude the conversation with Jonathan with the words 'You should definitely leave your wife', her answer would still be too hard and too direct. While it could lead to a deepened understanding of the meaning of Jonathan's alternatives, it would still completely ignore the *delicacy* (rather than the importance or the difficulty) of the question. That is what makes the word 'definitely' so inappropriate, even insulting here. This is very different from what we said with respect to the economist's answer. Adding the word 'definitely' adds to the acceptability of the answer 'You should definitely buy shares in company x'. Alexander may think: 'Well, if she is so sure about it, I should buy these shares immediately.' Mary's stressing the word 'definitely', however, would make her answer rather less acceptable to Jonathan, who is likely to think that Mary simply *cannot* be so entirely certain. He may even distrust her for her being so sure, thinking something like 'If Mary is so sure, maybe she just *wants* me to leave my wife. After all, she has always been very fond of me.' 'Definitely' thus emphasizes the harshness of Mary's answer, but it would be wrong to believe that leaving 'definitely' out would be leaving the hardness out: 'You should leave your wife' may still be too strong, even between good friends (unless,

[12] One could argue that what accounts for our different reactions to S1 (normal) and S2 (absurd) has to do with the fact that Alexander's question in S1 presents two mutually exclusive options (one cannot both go straight and turn left), while the options presented in S2 are not mutually exclusive (one can be married and do research). Consider, in that regard, S4, where Alexander asks whether he should buy shares in company x or y. These options are not mutually exclusive (one can buy shares in company x and in company y). In a realistic scenario, the economist could ask him why he believes that he has to either buy shares in company x or in company y, and this question could lead Alexander to achieve a better understanding of these alternatives. However, such an understanding is optional. If Alexander would say 'I don't want to achieve a better understanding of the alternatives, I just want the economist to tell me what I should do', we could say 'Ah, then that's all right'. By contrast, we expect Jonathan to want to achieve a better understanding of the alternatives. Paraphrasing Wittgenstein, we could say that he *ought* to want that. Thus, the contrast between Alexander's and Jonathan's cases remains, even if one compares two scenarios in which the options are not mutually exclusive. Consequently, the difference between their cases cannot be adequately explained by referring to the fact that Alexander's options in S1 are mutually exclusive while Jonathan's in S2 are not.

of course, Jonathan's case was very clear, but then it is difficult to understand why Jonathan would be struggling with the question in the first place).[13]

7.3.3 *Jonathan's Reaction*

What, then, is so problematic about Mary's answer 'You should leave your wife'? My suggestion is that the answer is too closed: it leaves Jonathan with no choice. It is as if Mary has already made the decision for Jonathan, and that is something that he may not appreciate. If Jonathan is to be a morally responsible and autonomous person, he has to make the decision for himself.[14] That is why his reaction to immediately leave his wife, irrespective of the content of the answer and irrespective of his interlocutor, has appeared absurd throughout this chapter. He may ask for advice, of course, but he has to seriously think about it before he can act upon it. In a notably relevant passage, Gaita writes:

> If I seek your advice on which is the best route off the mountain, then the nature of what I do in asking for it, and what you do in giving it, is conditioned by the fact that I may hand the problem over to you completely (perhaps you are better at it, or perhaps I have lost my nerve), by the fact that I may consult manuals and by other such familiar facts. But if I must make a moral decision by Monday, I cannot come to you on Friday evening, plead that I have little time over the weekend to think about it, and ask you, a rational and informed agent and a professor of ethics to boot, to try to have a solution, or at least a range of options, no later than first thing on Monday morning. (Gaita 2004: 103)

Alexander's questions are questions which can be handed over to someone else: he is happy to immediately act upon the passerby's and the economist's answers. He hands the question over to somebody who knows more about it and he lets experts make the decision for him. Moreover, it does not really matter who answers the question, as long as it is an expert. If Ada does not know where Alexander's hotel is, she may call her husband, ask him about it and pass on the answer to Alexander. If Ada the economist has no time to answer Alexander's question, she may refer him to a colleague or to an assistant. But what if Mary were to approach another friend of Jonathan's, saying: 'Well, Jonathan has asked me this question, but I have no time, could you think about it instead?'

So Jonathan's question is Mary's to answer, but his decision is, as Gaita formulates it, *'non-accidentally and inescapably' his* (2004: 103). This is why an answer that comes too close to a decision, an answer that does not leave enough room for Jonathan to make his decision himself, like 'You should leave your wife', is unwelcome. It ignores the delicacy of the question, even if 'definitely' is nowhere to be

[13] This does not mean that 'You should (definitely) leave your wife' is always too strong. I readily admit that there are cases in which such an answer would be appropriate and that, as a reviewer pointed out, 'sometimes our friends know what would be right for us better than we do'.

[14] For different views on this point, see Anscombe (1981), Driver (2006), Hertzberg (2002), Hopkins (2007), Jones (1999), Rhees (1999).

found in the answer. It smells of arrogance, while Mary's response instead needs what Paul Johnston calls 'moral modesty', or the recognition that she cannot make the decision for Jonathan and that she should not tell Jonathan what to do (Johnston 1999: 87).[15] This need for moral modesty, indeed, this *moral demand* for moral modesty is conspicuously absent in most moral philosophy.[16]

To say that Jonathan's decision is inescapably his to make is not to say that the content of Mary's response is of no importance, so long as she takes the question seriously, shows commitment, etc. After all, Jonathan is trying to find out what he should do. What Mary says may of course well have an important influence on Jonathan's decision, but what will influence Jonathan is not so much Mary's conclusion as her reasons. Things are precisely the other way round in Alexander's case. Mary's reasons may show Jonathan new ways of seeing his situation, ways of understanding what it would mean to decide this or that. They may point out things that Jonathan overlooked or somehow forgot to take into consideration. To say that Jonathan's decision is ultimately his, though, is to say that the weight of these considerations, their relevance for Jonathan's ultimate decision, are not for Mary to fix.[17] If Mary tells him 'You will regret it if you do not leave her', this is something that Jonathan may decide to take into consideration. It may very well help him, but in no way does it amount to telling him what he should do. He might judge that it is true, but unimportant. However, if Mary was to say 'You should leave her, because you will regret it if you don't', she would not be morally modest, because she would not be allowing Jonathan enough room to speak for himself.[18] Things are different in Alexander's case: presumably, Alexander expects the economist to fix the relative weight of different considerations, not just to point out everything that he should take into consideration. That does not mean that the economist would be entirely unhelpful in only doing the latter. It only means that Alexander may well feel that there is something lacking, that the economist could have done more to help him. Of course, it does not mean that Alexander would always be fully satisfied with a

[15] One should recognize that Johnston's notion of moral modesty slightly differs from mine, but the differences are not important for our present purposes.

[16] For example, Uri Leibowitz writes that 'General moral advice is a statement of the following form: *perform* action *A* if and/or only if ψ' (Leibowitz 2009: 350). Notable exceptions to the absence of a demand for moral modesty in moral philosophy can be found in Gaita (2004: 92) ('It is sometimes presumptuous even to *think* that another person ought to do something, even though we are quite sure what we ought to do if we were in their situation.') and Phillips (1964: 39) ('I do not propose to give positive advice. A man's life may be at stake whatever you do. I am prepared to clarify the issues involved as I see them, but you must draw your own moral conclusion.'). Not only does Phillips recognize here that the decision is to be made by the person who asks a moral question, there is also a clear commitment to clarification of the meaning of the alternatives. The moral advice offered here is reminiscent of what Wittgenstein says about Rhees's example (Rhees 1965: 22–23). It is not surprising, in this respect, that some of Wittgenstein's remarks on ethics seem to support a case for moral modesty (see, for example, Waismann 1965:16 and Bouwsma 1986: 45). For a detailed discussion of moral modesty and its relation to Wittgenstein's ethical views, see De Mesel (2014).

[17] See, on this point, Rhees (1999: 50, 89).

[18] On the importance of speaking for oneself in moral matters, see De Mesel (2015).

narrow answer either. He may very well want to know the economist's reasons for advising as she does. But, as we have seen, the role of reasons seems to be different here from their role in Jonathan's case. Asking for the economist's reasons is *optional* for Alexander, and it would not be strange if he were not interested in them.

So although Jonathan is clearly trying to find out what he should do, my conclusion is that we should not assume that he wants other people to tell him what he should do, even if he asks them what he should do. He may only want others to help him make a decision.[19] That is why it is an option for Mary, after serious consideration, etc., to be morally modest, so that she can ultimately leave the decision up to Jonathan. Nevertheless, if she thinks that she should draw a conclusion, a good way to do it may be to speak for herself. She could considerably soften 'You should leave your wife' by saying 'I would leave her' or 'If I were you, I would leave her'. Such an answer, evidently far more delicate than the rude 'You should definitely leave your wife', respects Jonathan's moral autonomy, stresses that Mary can only speak for herself, but at the same time (if Mary has seriously considered, etc.) is likely to help Jonathan in making his decision. Many other possibly helpful responses can be thought of, while in Alexander's case it is quite clear that the most helpful responses will be of the narrow answer form 'You should do x'. Suppose, for example, that Mary concludes her answer by saying 'I am terribly sorry, but I really cannot help you with this. I see no solution. But let's talk about it again soon.' It is my contention that such an answer, which is also a recognition of Jonathan's moral autonomy by someone who he knows and considers morally competent, could be the core of an ideal answer to his question. It shows that Mary understands the kind of question that Jonathan is asking her. She recognizes that the problem is delicate and fragile, that it requires a certain attitude and careful attention.

I do not want to suggest that in *every* case wherein a person asks for moral advice, it is appropriate for the advisor to tell the advisee that she sees no solution or to give a broad answer. In many cases, it would be paternalistic or even arrogant for an advisor to withhold her ideas about what a solution could look like. Again, I want to emphasize here what I wrote in Sect. 7.3.2: it is not my aim to argue that narrow answers are *never* appropriate, I only want to make a case for the much weaker claim that they are *not always* appropriate.

7.4 Moral Questions and Narrow Answers

We have seen in Sect. 7.2 that the difference between Jonathan's and Alexander's scenarios cannot be adequately explained by referring to a lack of knowledge or reasoning capacities of the passer-by, the probable unreliability of her narrow

[19] I agree with Rhees: 'Often I should like to know what so and so would have done in my place. And I may find it helpful or believe I shall find it helpful if he will tell me. Helpful in coming to a decision. This is not the same as looking to him for the answer' (Rhees 1999: 69).

answer or the difficulty or importance of the question. Rather, as Sect. 7.3 suggests, a question such as Jonathan's, in contrast to Alexander's:

(1) Is typically asked in a situation in which both parties in the conversation know each other personally, have a shared history;
(2) Asks for serious consideration, engagement, commitment and interest;
(3) Asks for an answer recognizing its delicacy;
(4) Asks for an answer offering a deeper understanding of the meaning of the alternatives;
(5) Asks for an answer recognizing that Jonathan's decision is non-accidentally and inescapably his.

These five characteristics are not exclusively tied to Jonathan's question. They are obviously present in many moral questions and we may assume that morally competent persons understand that they are. If Mary answers Jonathan's question by saying 'You should definitely leave your wife', she has misunderstood the kind of question that Jonathan has asked her and the kind of answer that Jonathan has asked her for.

One could object to my calling the five aforementioned points 'characteristics' of Jonathan's and similar questions. After all, they do not seem to characterize the question itself, but rather the situation in which it is asked. According to Wittgenstein, however, this dichotomy is deeply mistaken. To take a question out of the home in which it has its life, to ask it outside a particular language-game in order to study it, as philosophers tend to do, is likely to produce seriously distorted and ultimately useless results. It is as if one wanted to study the behaviour of a mammal in a vacuum. If one wants to understand what a question means, one has to look at the situations in which it is being used. What characterizes a question is what typically happens before and after, its typical context or 'surrounding', the circumstances in which we use it, what occasions it and what inferences are being drawn from it, what its typical antecedents and consequences are, what we use it for and what we do with it, how one typically reacts or responds to it, etc.[20] These are not external features, somehow added to the core meaning of the question. They determine what the question means and what kind of question it is in the first place. If one does not know the typical context of a question, what its antecedents and consequences are, how one typically responds to it, etc., one simply does not understand the question.[21]

---

[20] See Wittgenstein 1969: 103; Wittgenstein 1966: 68–72; Wittgenstein 1980a: §34, §571, §627, §780, §850; Wittgenstein 1980b: §149, §345.

[21] This way of putting things may be thought to obscure the distinction between semantics and pragmatics. If we want to know what a question means, then why would we have to look at the circumstances in which we use it or how one typically responds to it? The dichotomy between semantics and pragmatics in this objection, between what a question means ('in itself', as it were) and what its typical antecedents and consequences are, is deeply mistaken from a Wittgensteinian perspective. As Avner Baz remarks, philosophers looking for the meaning of something often commit the following mistake: '[…] differences that may be philosophically important between different words and the ways they function are either ignored, or else relegated to the realm of "(mere)

I do not claim that these five characteristics sharply distinguish moral questions from other kinds of questions, that these characteristics make questions moral questions or that they account for 'the' specificity of moral questions. I simply do not think that a list of necessary and sufficient characteristics, conditions or requirements could ever do so.[22] The boundaries between moral and non-moral questions are blurred and I will let that be so. Rather, I have tried to explain why S2 is absurd compared to S1 by reference to some characteristics of the question in S2. The characteristics of Jonathan's question allow us to say that *one cannot presuppose that moral questions of the form 'Should I do x or y?' ask for narrow answers, because broad answers will often do better in respecting the typical surroundings (partly reflected in these characteristics) of such questions.* As we have seen, broad answers have an advantage over narrow ones when it comes to, for example, offering a deeper understanding of the meaning of the alternatives or in recognizing Jonathan's moral autonomy. That is not to say that non-moral questions always ask for narrow answers, nor is it to say that moral questions never do.[23] (I readily admit that there are cases in which a narrow answer is what is asked for, indeed, in which it would be absurd to withhold a quick, straightforward moral judgment of the form 'You should do x'.) It is to say, however, that the demand for a certain kind of answer, recognizing for example the delicacy of the question, is not a demand for politeness. It is not a demand to couch one's answer in softer terms to achieve a certain effect, not a demand to add something to or do something with the essentially narrow

---

pragmatics" and taken to be inessential to their meaning' (Baz 2012: 18). If meaning is use, a sharp distinction between semantics and pragmatics will in many cases be unhelpful. However, that does not mean that the distinction becomes entirely useless. For a good discussion, see Baz (2012: 121).

[22] Driver (2006) shows that considerations of autonomy may require us not to give a narrow answer in moral as well as in non-moral cases. An anonymous reviewer suggests considering Alexander's money question in this regard: could it not be that considerations of autonomy will, in certain circumstances, require the economist to give a broad answer here? The answer is yes. What this shows is precisely that the characteristics of moral questions are not to be thought of as necessary (because considerations of autonomy are not equally relevant in all moral cases and may be even irrelevant in some) or sufficient (because considerations of autonomy may well be relevant in non-moral cases too) conditions and that the boundaries between moral and non-moral questions are blurred (one could argue that, by taking Alexander's autonomy into account, the advisor shows that she is sensitive to a moral aspect of what seemed to be a non-moral question). That does not, however, harm the overall argument. The fact that there are non-moral cases (or cases on the border between moral and non-moral) in which a broad answer is asked for does not weaken the point that it should not be assumed that moral questions ask for narrow answers.

[23] Some may find it misleading that I have been assuming that moral questions, although they do not always ask for narrow answers, are to be thought of as real questions. The early Wittgenstein, for example, wrote that 'a question [can] only [exist] where there is an answer' (Wittgenstein 2005: §6.51), and he seems to use 'answer' in the sense of 'narrow answer'. Thus, one could say that a question is not *really* a question if it does not ask for a narrow answer, or that it is then 'only' a broad question. However, this is not really an objection to my view. If someone denied that moral questions are always real questions because they do not always ask for narrow answers, I would be happy to allow for the term 'request' or 'invitation' instead of 'question'. I would be equally happy with 'response' instead of 'broad answer'.

answer to make it more acceptable.[24] And it is to say that it would be inconsiderate to simply assume the ideal answer to a moral question to be a narrow answer or that the ideal answer at its core was a narrow answer. It suggests that, in many cases, such as Jonathan's, a narrow answer is simply not what is being asked for by a moral question. One could take this to be reflected in Wittgenstein's remarks that 'ethics cannot be expressed' and that 'men to whom after long doubting the sense of life became clear, could not then say wherein this sense consisted' (Wittgenstein 2005: §6.421 and §6.521). In the *Tractatus,* Wittgenstein takes bipolar propositions as his model for what can be said. Because the answers to moral questions often cannot be cast in the mold of this model, the mold of a bipolar narrow answer of the form 'You should (not) do x', they do not belong to what can be said. However, there are ways of showing what cannot be said, of showing one's interlocutor some possible ways of thinking about or perceiving the situation. According to a strictly limited notion of 'saying', such as the early Wittgenstein's, showing these ways does not amount to saying anything. According to a narrow notion of 'answer', showing these ways is not answering the question, but dealing with it or treating it.[25] Barry Stroud writes: 'Treating a question is not the same as answering it. Answering it might be the worst thing to do with it. I believe that happens in philosophy' (Stroud 2001: 42).

Why is it important to note that moral questions do not always ask for narrow answers? Because exactly this assumption marks most philosophical accounts of morality, for example many of those that presuppose the ideal answer to a moral question to be a moral judgment.[26] Some Wittgensteinian philosophers, such as

[24] This is what Leibowitz suggests. Although he takes the general form of moral advice to be 'perform action *A* if and/or only if ψ' or 'perform action *A*', he admits that 'we sometimes give/receive advice in different forms; e.g., "think about the consequences", or "if I were you, I would do *A*". But strictly speaking, these forms of advice do not help us decide which action to perform; they could be helpful if they are understood as shorthand for something like the following: "think about the consequences, and perform the action that you believe would lead to the best possible consequences"; and "perform action *A* if you want to perform that action that I would have performed if I were you". So even if moral advice does not explicitly take the form: *perform* action *A* if and/or only if ψ, we should be able to restate it in this format' (Leibowitz 2009: 350).

[25] In my view, then, the conflict between the early Wittgenstein's 'ethics cannot be expressed' and the fact that the later Wittgenstein seems to allow for (or at least not to offer principled reasons against) meaningful moral language is not in the first place a conflict between views on ethics, but a conflict between understandings of 'saying' and 'expressing'. *If* one limits what can be said or expressed to bipolar propositions (as Wittgenstein does in the *Tractatus*), if one uses 'saying' and 'expressing' in a limited technical sense, *then* ethics cannot be expressed. But *if* one broadens what can be said or expressed to what we take 'saying' and 'expressing' to mean in everyday use (Wittgenstein 2009: §116), then ethics can be expressed. Thus, both the early and the later Wittgenstein would refuse to limit ethics to bipolar propositions, but neither the early nor the later Wittgenstein would claim that, if we understand 'saying' and 'expressing' as we do in everyday life, ethics cannot be expressed.

[26] Moral judgments are mostly understood as judgments that apply moral concepts to actions, persons or personal qualities (Crary 2007: 1), and that is how I use the term here. Although there is considerable disagreement over which concepts are moral, concepts like 'right', 'good', 'bad', 'wrong', 'duty', as well as concepts for virtues and vices ('courage', 'brutality', 'kindness') are commonly taken to be such concepts. To morally judge is to tell somebody what would be right,

Alice Crary, have convincingly shown that 'moral philosophers generally agree in assuming that moral thought invariably comes in the form of moral judgments' (Crary 2007: 1) and expressed 'hostility to the fixation on moral judgments characteristic of recent work in ethics' (Crary 2007: 3).[27] I agree with them on that point: a fixation on moral judgments is to be deplored, because it testifies to a misunderstanding of what many moral questions ask for. If only moral judgments are candidates for answers to moral questions, many broad answers that are not moral judgments (such as 'I'm terribly sorry, but I cannot help you, let's talk about it again soon', 'I don't know what to say', 'Maybe you could see it like this', 'You will regret it if you do not leave her' and 'You will save many lives') will simply not be taken into account. To focus on moral judgments is to unnecessarily narrow the range of possible answers to moral questions. It is not surprising, then, that those who only take moral judgments into account as possible answers to questions for moral advice tend to limit the range of possible answers to such questions to narrow answers (see, for example, Gibbons 2010; Kiesewetter 2011; Leibowitz 2009; Thomson 2001).

The fact that moral questions do not always ask for narrow answers is difficult (but not necessarily impossible) to deal with for moral theories. Most moral theories start from the fact that 'there is' a moral question, for instance the question 'Should Jonathan do x or y?', regardless of whether someone actually asks the question. They do not take into account the concrete dialogical or conversational situation in which the question is asked and has to be answered.[28] Given the characteristics that we mentioned, which are all tied precisely to this situation (they all refer to a dialogical situation and to the parties in that situation), this means that they do not understand the questions that they are trying to answer. Theorists then assume that questions such as Jonathan's require a narrow answer, like Alexander's questions do. Consequently, their dealing with Jonathan's question yields either:

(1) An answer of the kind 'Jonathan should do x', 'Jonathan should do y', 'Jonathan should not do x', 'Jonathan should not do y', 'Jonathan should do both' or 'Jonathan should not do either';
(2) The conclusion that the right narrow answer has not yet been found;
(3) The conclusion that there is no such determinate answer, that Jonathan faces a moral dilemma, meaning that Jonathan's doing x would be equally right or wrong as his doing y.[29]

---

good, bad, wrong, courageous, etc. for him/her to do, and therefore (if we assume that one has to do what is right, good or courageous and refrain from doing what is bad or wrong) to tell him/her what (s)he is obliged, should or ought to do.

[27] On this point, see also Diamond (1996 and 2002). For a detailed discussion, see De Mesel (2014).

[28] For example, theories claiming to provide a decision procedure often do not distinguish between the person who asks a moral question and has to make a decision (Jonathan) and the person who is asked to answer the question (Mary). The latter is offered a decision procedure, but she is not the one who has to make the decision. For Wittgenstein's insistence on the essentially relational character of ethical reflection, see Christensen (2011).

[29] For a useful overview of literature on moral dilemmas, see McConnell (2014).

In all cases, the ideal of a narrow answer is presupposed. The disagreement between moral theories is a disagreement over the content of our narrow answer, not a disagreement over the question of whether we should (try to) narrowly answer the question at all. The second case, in which no such answer has been found, is then seen by theorists as a failure on their part, while it is often thought to count against moral theories if they yield many conclusions of the third kind.[30] The problem, though, is not that no narrow answer is forthcoming, the problem is not that Jonathan's is a genuine dilemma, the problem is not that there are situations in which both answers are wrong, the problem is not the 'lack' of a determinate answer. The problem is that the theorist's ideal of a narrow answer, of a moral judgment, does not correspond to what both Jonathan and Mary, people actually dealing with a moral question, could reasonably regard as a good, appropriate, indeed as an ideal answer to Jonathan's question.

Some theorists may claim that there is no problem here; that the theorist is simply not looking for an answer that Jonathan and Mary could reasonably regard as good or appropriate. Of course, theorists have the right not to care about Jonathan and Mary, but it is difficult to see how, if they do not care, moral theory can claim to describe morality and/or offer practical guidance, which is what most moral theories do.[31] A description of morality involving a distorted notion of what a moral question is can hardly be called a good description. Practical guidance in the form of a narrow answer is bad practical guidance if a narrow answer is not appropriate in practice. I agree with Bernard Gert that 'any moral system that is proposed by philosophers will be judged by how closely that system coincides with the moral intuitions that thoughtful people have' (Gert 2005: 380) and with Uri Leibowitz (although he has a different notion of moral advice than mine) that 'If *S* is good moral advice, our moral theory should be able to explain this fact' (Leibowitz 2009: 352). If one accepts that, then there certainly *is* a problem if the theorist's ideal of a narrow answer does not correspond to what Jonathan and Mary could reasonably regard as a good answer to Jonathan's question.

The problem is that the theorist tends to misunderstand the question. His assumption that the question asks for a narrow answer may encourage Mary, if she studies moral theories, to be morally arrogant. Gert's notion of moral arrogance differs from mine, but it holds equally for my notion of moral arrogance that 'the ethical theories they [moral theorists] are putting forward might sometimes encourage those who read their theories to exhibit moral arrogance' (Gert 2005: 368). A first step towards moral modesty would be to take the first sentence of G.E. Moore's *Principia Ethica* seriously:

> It appears to me that in Ethics, as in all other philosophical studies, the difficulties and disagreements, of which its history is full, are mainly due to a very simple cause: namely to the attempt to answer questions, without first discovering precisely *what* question it is which you desire to answer. (Moore 1993: 33)

[30] This is why 'ethicists as diverse as Kant, Mill, and Ross have assumed that an adequate moral theory should not allow for the possibility of genuine moral dilemmas' (McConnell 2014).

[31] For an overview of literature on this dual role of moral theories, see Leibowitz (2009).

References

Anscombe, G.E.M. 1981. Authority in Morals. In The Collected Philosophical Papers of G.E.M. Anscombe. In *Ethics, Religion And Politics*, vol. 3, 43–50. Oxford: Blackwell.

Arendt, Hannah. 1958. *The Human Condition*. Chicago: The University of Chicago Press.

Baz, Avner. 2012. *When Words Are Called For. A Defense of Ordinary Language Philosophy*. Cambridge, MA: Harvard University Press.

Bouwsma, O.K. 1986. *Wittgenstein. Conversations 1949–1951*. Indianapolis: Hackett.

Christensen, Anne-Marie S. 2011. A Glorious Sun and a Bad Person'. Wittgenstein, Ethical Reflection and the Other. *Philosophia: Philosophical Quarterly of Israel* 39: 207–223.

Crary, Alice. 2007. *Beyond Moral Judgment*. Cambridge, MA: Harvard University Press.

De Mesel, Benjamin. 2014. Moral Modesty, Moral Judgment and Moral Advice. A Wittgensteinian Approach. *International Journal of Philosophy and Theology* 75: 20–37.

———. 2015. Speaking for Oneself. Wittgenstein, Nabokov and Sartre on How (Not) to be a Philistine. *Philosophy* 90: 555–580.

Diamond, Cora. 1991. Missing the Adventure. Reply to Martha Nussbaum. In *The Realistic Spirit. Wittgenstein, Philosophy, and the Mind*, 309–318. Cambridge, MA: MIT Press.

———. 1996. Wittgenstein, Mathematics, and Ethics. Resisting the Attractions of Realism. In *The Cambridge Companion to Wittgenstein*, ed. Hans Sluga and David G. Stern, 226–260. Cambridge: Cambridge University Press.

———. 2002. What if x isn't the Number of Sheep?' Wittgenstein and Thought-Experiments in Ethics. *Philosophical Papers* 31: 227–250.

Driver, Julia. 2006. Autonomy and the Asymmetry Problem for Moral Expertise. *Philosophical Studies* 128: 619–644.

Gaita, Raimond. 2004. *Good and Evil. An Absolute Conception*. 2nd ed. London: Routledge.

Gert, Bernard. 2005. Moral Arrogance and Moral Theories. *Philosophical Issues* 15: 368–385.

Gibbons, John. 2010. Things That Make Things Reasonable. *Philosophy and Phenomenological Research* 81: 335–361.

Hertzberg, Lars. 2002. Moral Escapism and Applied Ethics. *Philosophical Papers* 31: 251–270.

Hopkins, Robert. 2007. What Is Wrong with Moral Testimony? *Philosophy and Phenomenological Research* 74: 611–634.

Johnston, Paul. 1999. *The Contradictions of Modern Moral Philosophy. Ethics after Wittgenstein*. London: Routledge.

Jones, Karen. 1999. Second-Hand Moral Knowledge. *The Journal of Philosophy* 96: 55–78.

Kiesewetter, Benjamin. 2011. Ought' and the Perspective of the Agent. *Journal of Ethics & Social Philosophy* 5: 1–24.

Leibowitz, Uri. 2009. Moral Advice and Moral Theory. *Philosophical Studies* 146: 349–359.

McConnell, Terrance. 2014. Moral Dilemmas. In *The Stanford Encyclopedia of Philosophy*, ed. Edward Zalta. Spring 2014 Edition. http://plato.stanford.edu/entries/moral-dilemmas/

McGrath, Sarah. 2009. The Puzzle of Pure Moral Deference. *Philosophical Perspectives* 23: 321–344.

Moore, G.E. 1993. *Principia Ethica*. Rev ed. Cambridge: Cambridge University Press.

Phillips, D.Z. 1964. The Possibilities of Moral Advice. *Analysis* 25: 37–41.

Rhees, Rush. 1965. Some Developments in Wittgenstein's View of Ethics. *The Philosophical Review* 74: 17–26.

———. 1999. *Moral Questions*. Basingstoke: Macmillan.

Soanes, Catherine. 2006. In *Oxford Dictionary of English*, ed. Angus Stevenson. Oxford: Oxford University Press.

Stroud, Barry. 2001. What is Philosophy? In *What Is Philosophy?* ed. C.P. Ragland and Sarah Heidt, 25–46. New Haven: Yale University Press.

Thomson, Judith Jarvis. 2001. *Goodness & Advice*. Princeton: Princeton University Press.

Waismann, Friedrich. 1965. Notes on Talks with Wittgenstein. *The Philosophical Review* 74: 12–16.

Williams, Bernard. 1995. Who Needs Ethical Knowledge? In *Making Sense of Humanity and Other Philosophical Papers 1982–1993*, 203–212. Cambridge: Cambridge University Press.

Wittgenstein, Ludwig. 1966. *Lectures and Conversations on Aesthetics, Psychology and Religious Belief*. Oxford: Blackwell.

———. 1969. *The Blue and Brown Books. Preliminary Studies for the 'Philosophical Investigations*. 2nd ed. Oxford: Blackwell.

———. 1980a. *Remarks on the Philosophy of Psychology. Volume 1*. Oxford: Blackwell.

———. 1980b. *Remarks on the Philosophy of Psychology. Volume 2*. Oxford: Blackwell.

———. 1992. *Last Writings on the Philosophy of Psychology. Volume 2. The Inner and the Outer*. Oxford: Blackwell.

———. 2005. *Tractatus Logico-Philosophicus*. Trans. C.K. Ogden. London: Routledge.

———. 2009. *Philosophical Investigations*. 4ht ed., Trans. G.E.M. Anscombe, P.M.S Hacker, and Joachim Schulte. Oxford: Wiley-Blackwell.

———. 2014. A Lecture on Ethics. In *Lecture on Ethics*, ed. Edoardo Zamuner, Ermelinda Valentina Di Lascio, and D.K. Levy, 42–51. Oxford: Wiley-Blackwell.

Chapter 8
How Morality Can Be Absent from Moral Arguments

8.1 Introduction

What is a moral argument? A straightforward answer is that a moral argument is an argument dealing with moral issues, such as the permissibility of killing in certain circumstances. Let us call this the thin sense of 'moral argument'. Arguments that we find in normative and applied ethics are almost invariably moral in this sense. However, they often fail to be moral in other respects. In this chapter, I will discuss four ways in which morality can be absent from moral arguments in the thin sense. If these arguments suffer from an absence of morality in at least one of these ways, they are not moral arguments in, what I will call, the thick sense of 'moral argument'.[1] A moral argument in the thick sense is an argument that could possibly qualify as a proper response to a moral problem, an argument of which we can imagine that it would help someone struggling with a real moral problem (as a problem in life rather than as a problem in philosophy) to cope with that problem. If an argument in moral philosophy is not thick, then we should not let it weigh upon our considerations of what we morally ought to do, even if we see no independent reason to question the truth of the premises or the logical validity of the argument. The absence of morality in thin arguments means that these arguments will fail to give us a reason to do whatever they claim that we ought to do.[2]

How can morality be absent from thin moral arguments? I will start, in Sect. 8.2, with a few remarks made by Hans-Johann Glock about an argument of Peter Singer. I will argue that, if the conclusion of a moral argument involves or leads to the

[1] My distinction between 'thin' and 'thick' is related to, but different from the distinctions between 'thin' and 'thick' in Gilbert Ryle (2009, originally published in 1968), Clifford Geertz (1973), Bernard Williams (1985) and Michael Walzer (1994).

[2] I would like to thank an anonymous reviewer for the suggestion to put the point in these terms.

De Mesel Benjamin. 2016. How Morality Can Be Absent from Moral Arguments. *Argumentation* 30: 443–463. Published by Springer. See https://link.springer.com/article/10.1007/s10503-015-9389-8

B. De Mesel, *The Later Wittgenstein and Moral Philosophy*, Nordic Wittgenstein Studies 4, https://doi.org/10.1007/978-3-319-97619-8_8

denial of a moral certainty, then the argument is not moral in the thick sense and could not qualify as a proper response to a moral problem. The idea of a moral certainty used in this section is based upon an extrapolation of Wittgenstein's ideas in *On Certainty.* In Sect. 8.3, I will use the work of P.F. Strawson and Cora Diamond to support the claim that thin arguments often fail to recognize that a participant attitude is fundamental to morality. In Sect. 8.4, drawing on remarks and distinctions by Stanley Cavell, Cora Diamond and Raimond Gaita, I will point out a problematic reduction in many thin arguments of moral thinking to thinking about what to do. The reduction is problematic when it obscures the fact that moral thinking aims at understanding the meaning of what we do. In Sect. 8.5, I will show how the failure of certain thin arguments to take the personal character of moral problems into account disqualifies them as proper responses to these problems. In Sect. 8.6, the concluding, section, I will say something about how the distinction between thin and thick moral arguments invites a distinction between moral issues on one hand and moral problems on the other.

The aim of this chapter can be summarized by saying that I will offer a non-exhaustive overview of moral (as distinct from logical) reasons for not letting merely thin moral arguments weigh upon our considerations of what we morally ought to do. These reasons are not new. They are inspired by what I see as a Wittgensteinian tradition of moral philosophizing, since most of the authors discussed are heavily influenced by Wittgenstein.[3] What is new in this chapter is that (1) these reasons (and their authors) are brought together in such a way that an overview of overlapping features emerges, that (2) they are explicitly characterized, within the framework of thin and thick moral arguments, *as* reasons for not letting thin moral arguments weigh upon our considerations of what we morally ought to do. (3) They are not identified as 'logical' reasons (available irrespective of one's moral outlook), but as moral ones, meaning that their availability is tied to a specific moral outlook, an understanding of morality and/or of the possibilities and limitations of moral philosophy. The chapter aims to make explicit what kind of understanding or outlook is required for these reasons to be available. (4) Certain relations between these reasons and certain distinctions 'within' them (such as the different ways in which a thin moral argument may fail to account for the personal character of moral problems) will become clearer.

8.2 Denying a Moral Certainty

In *What Is Analytic Philosophy?*, Hans-Johann Glock remarks that 'ever since Plato, philosophers have shown an uncanny willingness to follow the argument wherever it leads' (Glock 2008: 194). Glock explains what he finds uncanny:

[3] I am speaking of a tradition in the sense explained by Glock, as 'held together *both* by ties of influence *and* by a family of partially overlapping features' (Glock 2008: 223). Glock distinguishes a tradition from a school, the latter being 'a tightly knit group based on relatively intimate personal contact and a direct transfer of certain doctrines of methods' (Glock 2008: 220).

> Even on reaching absurd or repugnant conclusions, they [philosophers] have rarely engaged in soul-searching or questioned their own premises. Instead, they have devised clever arguments for dismissing the judgements, values and practices of ordinary mortals as unreflective and obsolete. (Glock 2008: 194)

Glock uses Singer's *Practical Ethics* as an example. In the first edition of his book, Singer, according to Glock,

> [...] condones active *non-voluntary euthanasia*, the killing of innocent human beings that are incapable of understanding or making the choice between life and death – such as severely defective infants or grown-ups in a vegetative state. Moreover, he favours such a course of action not just in cases in which it is in the interest of the patient, but also in cases in which it is best for *the patient's environment* – the family or society. This includes both infants with Down's syndrome and haemophiliacs. (Glock 2008: 197)

Singer's position, Glock maintains, is 'a showcase for a particular failure of rationality [...]: the failure to reconsider one's premises in the light of unpalatable consequences, and the tendency to seek refuge instead in self-serving animadversions against 'orthodox' or 'conventional' morality and 'lay' intuitions' (Glock 2008: 198).

Glock does not, I think, at least not at first, see Singer's argument as logically invalid or his conclusion as, logically speaking, inadequately supported. Rather, he seems to think that the conclusion *cannot* be the right one. He refuses to think about it as something that could possibly be right. Not only Singer's argument, but *any* argument (regardless of its 'argumentative' quality) that leads to such a conclusion *must* be dismissed, precisely *because* it leads to that conclusion. The possibility of a sound argument for non-voluntary euthanasia is excluded. But what gives Glock the right to do that?

Nobody will dispute that, whenever an argument the validity of which we have at first sight no reason to question leads to an unexpected or controversial conclusion that goes against received views or intuitions, there is good reason to question its premises. However, its being unexpected or controversial does not imply that the conclusion cannot be right or that the argument cannot be sound. If it implied that, then there would be no room for criticizing received views. So if we want to hold on to the idea that certain conclusions, such as Singer's, cannot be right, then there has to be something more to them than their just being unexpected or controversial. Glock thus invites us to distinguish between, on one hand, arguments and conclusions that we are willing to evaluate because they could possibly be right, however unexpected or controversial they may be, and, on the other hand, arguments and conclusions that we refuse to evaluate because they could not possibly be right.

It is helpful, in this regard, to have a look at Wittgenstein's *On Certainty*. Wittgenstein discusses statements such as 'I have a brain' (Wittgenstein 1974: §4), 'My body has never disappeared and reappeared after an interval' (1974: §101) and 'The earth has existed long before my birth' (1974: §84). These statements are more aptly characterized as certainties than as beliefs. In contrast to beliefs, they cannot be meaningfully doubted or challenged (1974: §234). Wittgenstein considers the case of someone who goes to the doctor, shows him his hand and says 'This is a hand'. He then asks: '[...] if it is open to doubt "whether that is a hand", why isn't it also open to doubt whether I am a human being who is informing the doctor about this?' (1974: §460). These certainties belong to our 'frame of reference' (1974: §83), and to doubt them would be to 'knock from under my feet the ground on which I stand in making any judgments at all' (1974: §492). Certainties cannot be justified or denied because justifications or grounds for denying them will not be 'as certain as the very thing they were supposed to be grounds for' (1974: §307). Would we not refuse, asks Wittgenstein, 'to entertain any argument' that tried to show that the earth has existed for only a 100 years (1974: §577)? And would not such a refusal be reasonable? Here, 'we are not ready to let anything count as a disproof of this proposition' (1974: §245), and rightly so. Wittgenstein asks:

> What if it *seemed* to turn out that what until now has seemed immune to doubt was a false assumption? Would I react as I do when a belief has proved to be false? [...] Would I simply say 'I should never have thought it!' – or would I (have to) refuse to revise my judgment – because such a 'revision' would amount to annihilation of all yardsticks? (Wittgenstein 1974: §492)

What Glock finds uncanny about the philosopher's willingness to follow the argument wherever it leads is, I presume, that this willingness testifies to an unwillingness to consider Wittgenstein's question as a legitimate one. That is, because of an 'exaggerated confidence in the power of philosophical judgment' (Glock 2008: 194), philosophers often fail to see that there is an alternative to the 'I should never have thought it!'-option.

Wittgenstein only discusses (what we could, roughly speaking, call) empirical certainties, but some commentators have argued that it is plausible to extrapolate his account to the moral realm and to speak of moral certainties as well.[4] Nigel Pleasants, for instance, has convincingly argued that the wrongness of killing is such a certainty, and that it functioned as a certainty even for slaveholders and in the Third Reich (Pleasants 2008; Pleasants 2015). It should be remembered that Glock describes Singer's argument as condoning 'the *killing* of innocent human beings' (my italics). Using Wittgenstein's conceptual tools, we could say that one of Glock's reasons for refusing to accept Singer's argument as an argument that could possibly be right is that it leads to the denial of a moral certainty. Glock understands 'it is wrong to kill innocent human beings' as relevantly analogous to 'the earth has existed long before my birth'. According to Wittgenstein, it is 'absurd' to doubt, justify or deny certainties (Wittgenstein 1974: §460), and it is conspicuous that Glock uses the very same term 'absurd' to characterize the conclusions of the philosophers that he challenges (Glock 2008: 194).

According to this proposal of forming an analogy between 'empirical' and moral certainties, 'it is wrong to kill innocent human beings' would be a statement that cannot be meaningfully doubted or challenged. Thus, it belongs to our frame of reference and to doubt it would be to knock from under our feet the ground on which we stand in making any moral judgments at all. Grounds for denying it will never be as certain as the very thing that they are supposed to be grounds for and we are not ready to let anything count as a disproof of this proposition.

---

[4] For a recent overview of the discussion and helpful references, see Pleasants (2015).

Although the analogy between empirical and moral certainties may seem attractive at first sight, there is what can be called a problem of criticism. The distinction between beliefs and certainties saves the possibility of criticizing received beliefs, but it raises questions about the possibility of doubting or criticizing certainties. Wittgenstein's point is not that all certainties have remained the same throughout history and that they stand forever. It is not impossible that, in certain circumstances, what was immune to doubt at one point becomes open to doubt at another; what was certainty becomes belief and the other way round:

> The mythology may change back into a state of flux, the river-bed of thoughts may shift. But I distinguish between the movement of the waters on the river-bed and the shift of the bed itself; though there is not a sharp division of the one from the other. (Wittgenstein 1974: §97)

One could think here, I suppose, of the way in which Galilei made it possible to criticize what until then had functioned as a certainty, namely the certainty that the sun revolves around the earth. That certainty became a criticisable belief, the belief was shown to be false and it was replaced by the belief that the earth revolves around the sun, a belief which has now become a new certainty.

The question now is as follows: could we not regard Singer as someone who does for morality what Galilei did for astronomy, that is, as someone who makes it possible to criticize what many regard as immune to doubt? If the consequence of such criticism is that the whole frame of reference changes, then why should we bother about that? What matters is not whether the frame of reference changes or not, but whether it is the right frame of reference. If Galilei was right, it seems that Singer could be right as well.

Or not? Those who maintain, with Glock, that Singer's conclusion cannot be right, have resources to respond to this challenge. They could argue that moral certainties are different from at least some empirical certainties when it comes to the possibility of radical change. After all, Wittgenstein does not hold that *all* certainties are open to change:

> And the bank of that river consists partly of hard rock, subject to no alteration or only to an imperceptible one, partly of sand, which now in one place now in another gets washed away, or deposited. (Wittgenstein 1974: §99)

The question then is why moral certainties would be 'hard rock' (or more likely to be hard rock than empirical ones), while at least some empirical certainties are not. Sabina Lovibond suggests an answer (or the direction to an answer) to that question when she says that, in moral matters, we do not have the idea of *fresh evidence,* an idea that "belongs to' our concept of the physical world, but not to our (possible) concept of the moral world' (Lovibond 1983: 79–80). Although I suppose that fresh evidence could not possibly influence the hard rock certainty that this is my hand, for example, it has justified scientific revolutions and revolutions in our conception of the physical world. Similar revolutionary changes in our conception of the moral world would then be unjustifiable, because there is no such thing as fresh evidence or new discoveries here. Peter Hacker makes a related point about evidence and discoveries in relation to philosophy. If we give it a moral twist, it would go as follows: 'The characteristic reaction to an advance in scientific knowledge is 'Goodness me, who would have thought of that!', whereas the characteristic response to a moral [Hacker has 'philosophical'] insight is 'Of course, I should have thought of that!'' (Hacker 2009: 148). Hacker thus suggests that the 'I should never have thought it!'-reaction is not open to us in morality. The moral philosopher's task would thereby not be to revise or revolutionize, but rather to make us understand. Moral exemplars such as Gandhi and Nelson Mandela did not come up with new evidence or revolutionize our conception of what is morally advisable or permissible. Rather, they deepened or reminded us in a powerful way of what, in a sense, we already knew. As Raimond Gaita formulates it,

> Ethical understanding is often coming to see sense where we had not seen it before, or coming to see depth where we had not seen it before. It is seldom learning something completely new (there are no Nobel Prize-winning discoveries in ethics) and it is seldom seeing that there is, after all, a valid argument to support positions we had previously judged to be dubious. It is often seeing what someone has made of something that we had often heard before. (Gaita 2004: 281)

The line of response suggested by Lovibond, Hacker and Gaita does not lead to a conventionalist position and is not meant to prove that, in moral matters, no such thing as a revolution is possible. If we understand a moral revolution to be a radical change in our ways of dealing with and thinking about moral matters, then, for example, the fact that we no longer regard slavery as morally permissible shows that there has been a moral revolution. But in contrast to scientific revolutions, such a revolution is typically not the result of new discoveries or fresh evidence. Moreover, some proposed way of acting or thinking will only count as a candidate for a possible moral revolution if it somehow succeeds in showing that our moral certainties lead us into (or at least do not conflict with) acting or thinking in such ways, while something, if it is supported by evidence, might be a candidate for an empirical revolution even when it goes against empirical certainties (see the Galilei example). If a proposed way of acting or thinking is understood to go against the moral certainty that killing innocent people is wrong, for example, then we will typically not consider it to be a good candidate for a moral revolution, and rightly so. This is why Glock dismisses Singer's argument. If, on the other hand, a radically new proposed way of acting or thinking can be shown to be implied or suggested by our moral certainties, its prospects are much better. Take, for example, the idea that killing animals is wrong. The certainty that killing people is wrong, in combination with the belief that animals are relevantly similar to people (engendered, for example, by Frans de Waal's (2006) observations regarding the competences of certain animals), may very well lead one to think that it is wrong to kill animals. Such a belief seems like a good candidate for a moral revolution, precisely because it does *not* conflict with our moral certainties. Singer's critique of speciesism (Singer 2009) is therefore much more likely to spark a moral revolution than his argument in favour of non-voluntary euthanasia (at least if we understand the latter as Glock understands it, namely as condoning the killing of innocent human beings).

If a moral revolution is understood as a radical change in our dealing with and thinking about moral matters, then the idea of moral certainty does not exclude moral revolutions. If, however, a moral revolution is taken to be a change so radical that, for example, 'killing innocent people is wrong' would no longer be a certainty, then it is doubtful as to whether we can conceive of anything like a moral revolution at all. If people were to start randomly killing each other, then they have not so much radically changed their moral system or their morality. They should rather be said to have left the moral sphere completely. It would be hopelessly inadequate to call what these people are doing 'morally wrong', 'cruel' or 'immoral', that is, to condemn it in everyday moral terms. Rather, we would say that they do not know what morality is or is about. Their revolution is not a revolution from one moral way of thinking to another, it is a revolution from morality to something else, and it is at least doubtful whether we would still be able to see these people as living a human form of life (see, on this point, Rummens 2013: 146).[5]

Lovibond's and Gaita's remarks about morality and moral philosophy do not exclude moral revolutions. Rather, they show that those who maintain that Singer's conclusion cannot be right (because it involves or leads to the denial of a moral certainty) have resources to respond to those who (claim to) see no crucial difference between criticizing certain empirical certainties, such as the certainty that the sun revolves around the earth, and criticizing moral ones, such as the certainty that killing innocent people is wrong. It is morally and philosophically defensible to say that the conclusion of a moral argument cannot be right because it involves or leads to the denial of a moral certainty. By 'philosophically defensible', I mean that it does not suffice to assert, dogmatically, that a certain conclusion denies a moral certainty and should therefore be dismissed (as it does not suffice to assert, dogmatically, that everything is open to doubt). There are criteria for what counts as a certainty and those who defend that something is a moral certainty have to show that it fulfills these criteria.[6] Moreover, they will be committed to subscribe to a certain understanding of morality and moral philosophy (see Lovibond's and Gaita's under-

[5]A reviewer has remarked that it is not really clear where to draw the line between moral change or evolution and moral revolution. I cannot deal with the issue at length here, but I can say, first, that an answer to this remark crucially depends on one's understanding of 'moral revolution'. If a moral revolution is a change so radical that our moral certainties such as 'killing innocent people is wrong' would no longer be certainties, then it is doubtful, as I have tried to make clear, whether we can conceive of anything like a moral revolution at all. If, on the other hand, a moral revolution is a radical change in our dealing with and thinking about moral matters, then there can be moral revolutions, but whether something is a revolution or an evolution will then depend on how radical the change is. I would say that the way in which we think about slavery has undergone a revolution rather than an evolution, while the way in which we treat animals has undergone an evolution but not (yet?) a revolution. Admittedly, the distinction is not clear-cut, and neither is the distinction between 'radical' and 'not radical'. The fact that there is a grey zone here does not prove, in my view, that the distinction is useless, and I do not even know whether we should try to draw a clear line between evolutions and revolutions. See, on a related point, footnote 14.

[6]A good overview of the characteristics of certainties can be found in Rummens (2013). I will only briefly mention them here. (1) Basic certainties cannot be meaningfully doubted. (2) Basic certainties cannot be justified. (3) Basic certainties are certainties of our acting. (4) Doubt regarding basic certainties is a form of insanity. (5) Basic certainties are the preconditions of local doubt. (6) Basic certainties form a system. (7) Basic certainties are not necessarily certain (they are not conceptual or *a priori* truths) (Rummens 2013: 134).

standing of morality and moral philosophy). I conclude that, if a thin moral argument involves or leads to the denial of a moral certainty, and given a certain understanding of morality and moral philosophy, there is good reason to regard that thin argument as one that we should not take into consideration when deliberating about what we morally ought to do.

8.3 No 'Attitude Towards a Soul'

Another (but obviously related) reason for refusing to accept a moral argument as one that could possibly qualify as a proper response to a moral problem is the absence of a moral attitude in the way that the argument deals with the issue. Cavell discerns this kind of 'absence of morality' in Stevenson's *Ethics and Language* (Cavell 1999: 274–291). Cavell finds fault, in particular, with Stevenson's saying that '*Any* statement about *any* matter of fact which *any* speaker considers likely to alter attitudes may be adduced as a reason for or against an ethical judgment' (Cavell 1999: 274). According to Cavell, Stevenson commits a sort of category mistake and is unable to recognize that some reasons can be of the wrong kind.[7] Cavell accuses Stevenson of not seeing that some kinds of reasons do not have a place in thick moral arguments, even if a speaker considers them likely to alter attitudes. If we look back at Glock's description and dismissal of Singer's argument and conclusion, precisely this idea of reasons being of the wrong kind seems to play a major role. One of the reasons Singer invokes, according to Glock, for condoning the killing of certain human beings is that it is sometimes best for their environment. The problem with this reason is not so much that it does not support (or inadequately supports) the conclusion, but that it is the wrong kind of reason. Why?

Singer's reason, it could be said, testifies to what P. F. Strawson has called an 'objective attitude' (Strawson 2009: 155) towards other human beings. Such an attitude sees them as subjects for treatment or objects of social policy, as something 'to be managed or handled or cured or trained' (Strawson 2009: 155). An objective attitude 'cannot include the range of […] attitudes which belong to involvement or participation with others in inter-personal relationships' (Strawson 2009: 156). The objective attitude is opposed in Strawson's paper to a participant attitude, and I understand such a participant attitude to be relevantly similar to what Wittgenstein has called 'an attitude towards a soul' (Wittgenstein 2009b: §22).

The idea that a moral argument should embody a participant attitude in order to qualify as thick may be understood in three different ways, depending on the answer to the question 'a participant attitude towards whom?' The first way to answer this question is by saying that a thick argument should involve a participant attitude towards the persons that we are arguing *about*. Suppose that Sophie tries to convince John that he should bring his mother to a retirement home. Sophie's argument could only qualify as a proper response to John's moral problem if it embodies a

[7] I would like to thank an anonymous reviewer for the suggestion to put the point in these terms.

participant attitude towards John's mother. The second way to answer the question is by saying that a thick argument should involve a participant attitude towards the persons we are arguing *with.* So in the case of Sophie and John, Sophie's argument could only qualify as a proper response to John's moral problem if it embodies a participant attitude towards him. The third answer is that, in order for a moral argument to be thick, it should involve a participant attitude towards both those about whom we are arguing *and* those with whom we are arguing, and this is the way in which I propose to understand the idea that a thick moral argument should embody a participant attitude or an attitude towards a soul.[8] If such an attitude is lacking in a moral argument, we have good reason not to let it weigh on our deliberations about what we morally ought to do. The thin argument fails to give us a reason to do whatever it claims that we ought to do because it misses something that we regard as essential to morality and thus fails, in a thick sense, to be a moral argument at all. If what purports to be a response to a moral problem reflects a misunderstanding of morality, then why should we bother about what it says we morally ought to do?

Is it not dogmatic to argue that 'This argument fails to display a moral attitude, therefore it cannot be accepted'? As dogmatic, I would say, as 'This is a moral certainty, so it cannot be meaningfully denied'. That is, it does not suffice to assert that an argument does not embody a moral attitude. A characterization of that attitude and an explanation as to why one takes this attitude to be characteristic of or essential to morality are required. Again, the availability of reasons like 'this argument is not moral at all' (in the thick sense) or 'these are reasons of the wrong kind', will crucially depend upon one's own conception of morality. That is not, however, a weakness: everyone operates with *a* conception of morality. It seems much more dubious to claim that these reasons are available independently of one's conception of morality, for that would make them morally neutral, while those who invoke them *want* them to reflect a moral outlook. They do not reject the argument on logical grounds (available irrespective of one's understanding of morality or moral outlook), but on moral grounds.

I have focused on the absence of a participant attitude in moral arguments. Yet such an absence often occurs much 'earlier', namely, in the formulation or presentation of the problem. In 'What if x isn't the number of sheep? Wittgenstein and Thought-Experiments in Ethics' (2002), Cora Diamond refers to a discussion by Carol Gilligan of a hypothetical case in which a man must decide whether or not to steal the drug necessary to save his wife's life. When confronted with the dilemma,

---

[8] In an earlier version of this chapter, I maintained that a thick moral argument should embody what Lawrence Blum has called 'a sense of shared humanity, of regarding the other as a fellow human being' (Blum 1994: 177). Reviewers have remarked that this formulation raises questions about whether we should restrict the moral community to humans alone. I believe that 'a participant attitude' and 'an attitude towards a soul' are better ways of making the point, since these formulations do not exclude that we can take up these attitudes towards animals as well. This is not to say that we only have moral responsibilities towards other human beings and animals, and not, for example, towards the environment. The need for a participant attitude will boil down, in these cases, to the need for a participant attitude towards those we are arguing with, an attitude that shows awareness of our shared responsibility towards the environment.

a girl said that communication should be re-opened with the druggist, although it had been stipulated that no further communication was possible. One could respond here, as Roy Sorensen did, that the girl's response is misconceived because the man's situation 'isn't being reported […], it is being *stipulated*' (Diamond 2002: 238). Hence, according to Sorensen, 'it's no more open to say in response that communication should be re-opened with the druggist than it is to respond to the mathematical stipulation "Let x be the number of sheep" as did the schoolboy in the story, "But, Sir, what if x isn't the number of sheep?"' (Diamond 2002: 238). The problem with this response, Diamond says, is that 'in moral life we are not in general confronted with cases the description of which can be taken to be simply "given"' (Diamond 2002: 238). The girl's answer is one that rejects the presentation of the problem because that presentation already leaves out a the possibility of a participant attitude: it does not represent the druggist as a fellow human being, a being that could be responsive to our reasons or pleas for re-opening communication. Hence, the presentation of the problem already precludes what would be a moral course of action. According to Diamond, the girl has good moral reason to answer as she does.

Diamond's general point is that, in moral matters, 'describing the situation is frequently itself part of the problem' (Diamond 2002: 239). The point is illustrated by Elizabeth Anscombe's reaction to the use of hypothetical dilemmas by consequentialists 'who ask whether it would not be right to execute an innocent person if that were the only way to avert a nuclear war' (Diamond 2002: 239). According to Anscombe, 'the most important thing about the way such cases are invented in discussions is the assumption that only two courses are open' (Diamond 2002: 239). Diamond explains:

> […] it's no part of Anscombe's view of moral thought or moral theorizing that we should be able to say what it would be good to do in every stipulatable set of circumstances.[9] Her kind of moral thinking (and other kinds of moral thinking as well) rejects the invitation to resolve in advance, on the stipulated terms, horrific hypothetical dilemmas. One might hope to have, if one ever were in such horrific circumstances, the capacity to re-conceive the situation, to see possibilities at first unsuspected. […] He [Sorensen] obviously thinks Anscombe's suggestion about the nuclear war dilemma is simply an attempt to avoid the fact that thought-experiments like that one have embarrassing consequences for her kind of approach to ethics. But her view is rather that they have no consequences at all; they are simply part of a consequentialist conception of what a moral theory should do, and of what moral life is like. (Diamond 2002: 239–240)

Similar considerations are voiced by Gaita, who discusses the example of a person presented a with similar dilemma (Gaita 2004: 64–73). If he kills one innocent person, ten others will be saved. If he does not, they will all be killed. To present the matter in this way, Gaita claims, is to simplify the man's sense of what he might be doing. 'It is already to see the situation in the morally flat light of consequentialist

[9] According to Gaita, 'it is mere prejudice to believe that it is an obvious virtue of a philosophical account of ethics to characterise our sense of the ethical in such a way as to yield a decision procedure for what to do in any conceivable situation' (Gaita 2004: 73).

theory' (Gaita 2004: 71), a light, we can say, that prevents us from responding to the situation in a way that reflects an attitude towards a soul.

In short, if a thin moral argument presents a moral problem in a way that involves no participant attitude, or if the argument, the reasons provided for the conclusion or the conclusion itself, suffer from such an absence of morality, then that is a reason not to think of the argument as moral in the thick sense, given a certain conception of morality. Slightly adapting a phrase from Wittgenstein's 'Lecture on Ethics', we could say that 'When we look at it in this way, everything moral has disappeared' (Wittgenstein 2014: 50).

8.4 No Deepened Understanding

How can a thin moral argument fail to be thick? In this section, using distinctions by Cavell and Diamond, I will argue that thin arguments often present moral problems as well-posed problems, while it may be more appropriate to treat them as exploration problems.

Consider, first, the following dialogue, an example that runs through Cavell's *The Claim of Reason*:

A There is a goldfinch in the garden.
B How do you know?
A It has a red head.
B But goldcrests also have red heads.

B has opened ground for doubt and thereby reveals that A's claim about the goldfinch has been insufficiently supported. According to Cavell,

> [...] if the argument is to continue then either the ground for doubt must itself be impugned ('The shape of a goldcrest's head is different') or a new basis proposed ('I know not just from the head, but from the eye-markings'), and every person competent to enter claims to knowledge knows the significance of this. We could say: It is not up to the protagonists to assign their own significance to bases and grounds for doubt; what will count as an adequate basis and sufficient ground for doubt is *determined by the setting of the assessment itself.* When I counter a basis by saying 'But that's not enough', there is no room for you to say, 'For me it is enough'. (Cavell 1999: 267)

Cavell compares the goldfinch example to several moral cases. I propose to compare it to a new example:

A Barry is an untrustworthy person.
B Why do you believe that?
A He promised to talk about our problems yesterday evening, but he didn't turn up. He didn't even send a message and he didn't pick up the phone.
B But his mother fell ill 2 days ago. He is often with her, she counts on him.

As in the goldfinch example, the ground for doubt can be impugned ('I called his mother to ask if he was there. He was not.') or a new basis can be proposed ('It's not just about what happened yesterday. Last month, for example, he ...'). In contrast to the goldfinch example, however, there *is* room for A to refuse to accept B's ground for doubt, 'without impugning it as false, and without supplying a new basis, and yet not automatically be dismissed as irrational or morally incompetent' (Cavell 1999: 267). Although A, if she is morally competent, will not fail to see that B's consideration requires a determination by her, it is up to her to determine whether what B says is *enough* to counter her claim to be right or justified in saying that Barry is untrustworthy. A could say, for example, 'That does not make him more trustworthy', and it could be reasonable for her to say so. In the goldfinch example, by contrast, 'the relevance of the doubt is itself enough to impugn the basis as it stands' (Cavell 1999: 267).

What Cavell's distinction between the goldfinch example and moral cases tells us about moral arguments is, I take it, that what will count as an adequate basis or sufficient ground for doubt is often unclear and even in question in moral arguments, while it is often treated as given or taken for granted in other kinds of argument. Something similar holds for the meaning of moral concepts. While the meaning of a concept is often treated as given in arguments, in moral arguments this meaning is often unclear or in question (Murdoch 2001: 27). That does not mean that moral arguments are explicitly *about* sufficient bases, grounds for doubt or the meaning of moral concepts (they sometimes are, of course). Instead, it means that questioning these bases, grounds and meanings is a legitimate move within moral disputes, even if they are not explicitly concerned with bases, grounds and meanings. There is room, within moral arguments, to question bases, grounds and meanings, and this room characterizes moral arguments (which is not to say that it cannot be characteristic of other kinds of argument, such as arguments in aesthetics or arguments in science). If, for example, in response to a question about whether a given person ought to be punished, one simply refers to the rules of an institution (and thereby refuses to acknowledge that, within a moral argument, there is room to question these rules), then that is, according to Cavell, 'to refuse to allow a moral question to be raised' (Cavell 1999: 303).[10]

A similar point has been made by Wittgenstein in a discussion recorded by Rush Rhees. Rhees presents Wittgenstein with the example of a man who has come to the conclusion that he must either leave his wife or abandon his work of cancer research.[11] Wittgenstein says that 'here, we may say that we have all the materials of a tragedy' if we are talking about a man 'who does not have an ethics' (Rhees 1965: 23). But if he does have an ethics, his problem is different. If he is a Christian, for example, 'he has got to stick to her come what may. […] The question 'Should I leave her or not?' is not a problem here' (Rhees 1965: 23). So, a question or problem that can be solved (that one accepts to solve or treat) within a system of rules or by reference to an institution in which it is clear what adequate bases and sufficient ground for doubts are and what concepts mean, is different from a question or prob-

[10] Compare Murdoch: '[…] it is characteristic of morals that one cannot rest entirely at the conventional level, and that in some ways one ought not to' (Murdoch 2001: 29).

[11] Here and elsewhere in this chapter, the discussion resonates with and repeats some of the discussion of Chap. 7. I have chosen not to eliminate the overlap, so that it remains possible to read the chapters on their own.

lem that cannot be so solved (that one refuses to so solve or treat), a question or problem that invites exploration of and discussion about these meanings, bases and grounds. I will follow Diamond in calling the former kind of problems 'well-posed problems' and the latter 'exploration problems' (Diamond 2002: 241–242), although my distinction may be slightly different from hers.

We have seen that, if we call an argument 'moral' in virtue of the fact that it concerns moral issues (the permissibility of killing, for example), it can fail to be moral in the thick sense if it does not embody a participant attitude. In Sect. 8.2, we considered a question to be rightfully treated as well-posed, but the bases and grounds for doubt one invokes, and sometimes also the description of the problem, and the attitude towards the problem these reflect, are not those of (do not fit in) a moral (in the thick sense) system or institution. As the Cavell-Wittgenstein-Diamond distinction shows, a thin argument can also fail to be moral in the sense that it treats as well-posed a problem or question that is not obviously to be taken as well-posed (that could competently be taken as an exploration problem). If it does so, I argue, the thin argument will fail to give us reason to do whatever it claims that we ought to do.

Take Wittgenstein's example of the man who has come to the conclusion that he must either leave his wife or abandon his work of cancer research. Suppose that this man, Billy, turns to his best friend Lynn for help and presents her with the dilemma. She answers his request by offering him a moral argument. She sets out the premises and starts reasoning. Billy has no reason to think that her conclusion will involve the denial of a moral certainty, and the way she argues clearly testifies of a participant attitude. Lynn tries to help Billy to re-conceive the situation, to see possibilities at first unsuspected, and so on. Yet something disturbs him: she obviously intends to reach a conclusion and tell him what would be best for him to do. She treats his problem as if it were well-posed, his question as if it were asking for a solution of the form 'You ought to do x'. And that may be something Billy does not want and, independently of what Lynn's conclusion will be, something he can reasonably refuse to accept, because he sees his problem as an exploration problem or at least as having an ineliminable and central exploration aspect. The point is not that he is not trying to find out what he should do (he clearly is), but that, even if he asked her explicitly what he should do, he may only want her to help him make a decision (see, on this point, Rhees 1999: 69). He may think or feel that it is not up to her to determine whether the reasons she invokes for or against a certain way of acting are 'enough'. Lynn's response comes too close to a decision. It leaves no room for discussion about meanings, adequate bases and grounds for doubt, a kind of discussion that he may take his moral problem to have invited. It is closed in a way that moral arguments should not be. It is concerned exclusively with what Billy should do, whereas, as Gaita has insightfully remarked,

> [...] much moral thinking is not thinking *what to do,* and even when it is it is also an attempt to understand the *meaning* of what we do, which is rarely thinking about the empirical consequences of what we do, or about how our principles stand in relation to those consequences and to one another. It is, most often, an attempt to achieve a deepened understanding of the *meaning* of our actions. That, as much as conscientiousness, is a mark of moral

seriousness and of what we should understand by having arrived at a 'right answer' to a moral question. (Gaita 2004: 264)

Billy is likely to think that he cannot and should not take or accept a decision, that he cannot and should not answer or accept an answer to his question unless he understands as adequately and deeply as possible the meaning of his problem and of the alternative courses of action open to him. If Lynn's argument does not offer such an understanding, because it reduces moral thinking to thinking about what to do (ignoring the exploration part of his problem, taking for granted that what it presents as the meaning of certain moral concepts and as sufficient bases and grounds for doubt are not in dispute), it will fail to give Billy reason to do whatever it concludes that he ought to do. The problem is not that Lynn's answer is or cannot be the right one (as in Sect. 8.2), or that it lacks a participant attitude (as in Sect. 8.3 but that it is not an answer to his question or a solution to his problem. It misjudges the kind of problem that he has.[12]

8.5 Ignoring the Personal

I will argue that, if a thin moral argument ignores the personal character of the problem that it purports to deal with, we should not let it weigh upon our considerations of what we morally ought to do.[13] The question then is as follows: in what sense are moral problems rightfully thought to be personal, and how can thin moral arguments fail to capture this? I will distinguish between three ways in which a thin moral argument's failing to be personal gives us a reason to reject it as a proper response to a moral problem.

First, what people ought to do may reasonably be thought to depend upon what they are capable of doing (if 'ought' implies 'can'), on what their responsibilities are, what their history is, etc., in short, on their personal characteristics. According to Rhees, personal characteristics determine not only what someone ought to do, but also what her moral problem *is*. Reasons may weigh differently from one person to another, 'and what makes it the problem it is for me are the reasons which weigh with me in the one direction and in the other' (Rhees 1999: 50). If the weight of reasons makes a moral problem the kind of problem that it is (Billy's problem would not be the same problem, or not even a problem, if his reasons would not weigh with him as they do), and if different people weigh moral reasons differently and are justified in doing so (as the difference between the goldfinch case and Barry's case and Cavell's comments upon such cases suggest), then another person cannot solve my moral problem in the sense that an ornithologist can solve my goldfinch prob-

[12] See, on this point, Chap. 7.

[13] Note that I write 'personal' and not 'subjective'. The pairs personal-impersonal and subjective-objective are not equivalent, for the personal may be objective. See, on this point, Chap. 5. While many philosophers will be inclined to say that morality is personal in some sense or another, not so many hold that it is subjective.

lem. This raises well-known difficulties for issues of moral advice, moral expertise and moral deference. What is important for our purposes is that these problems go together with questions about the teachability of ethics and the possibility (and desirability) of action-guiding moral theory. While some just note that the individual or personal character of moral problems and their solutions makes it 'difficult to learn goodness from another person' (Murdoch 2001: 29), others argue that it makes teaching ethics impossible.

Such an outright dismissal of ethical theory and of the teachability of ethics may seem exaggerated. Even if moral problems are personal in the sense that reasons weigh differently with different persons, it may be argued that this does not harm the teachability of ethics. And even if the personal character of moral problems makes it misleading to talk of moral theories, it does not necessarily lead to problems for action-guiding moral arguments. The reason is that such arguments can take all relevant personal characteristics (among which the way reasons weigh with this or that person) into account. The question then simply changes from 'What ought to be done in this situation?' to 'What ought this person to do in this situation?' Billy, for example, can inform Lynn about how reasons weigh with him, and she can take this information into account when offering him a moral argument. Moral arguments, it seems, do not necessarily have difficulties with the personal character of moral problems.

Or do they? It should be noted that 'taking personal characteristics into account' is easier said than done. Billy can be mistaken or unsure (which seems rather common) about how reasons weigh with him, and Lynn will have to know him very well in order to judge that. If a person's unique history is often important for determining what he ought to do, then taking relevant personal characteristics into account will amount to taking 'the details of *this* personality' (Murdoch 2001: 22) into account. Generally speaking, we can say that the more that a moral argument succeeds in taking someone's unique personality into account, the more restricted the scope of the argument will be: it will, ultimately, only tell one person what (s)he ought to do in a specific situation. If that is what moral arguments (should) do, relevant instances will be found in literature and in people's personal lives rather than in moral philosophy. If, on the other hand, the argument is less personal, its scope will be broader. It will apply to what philosophers, attempting to steer a middle course between extreme particularism and crude, impersonal generalism all too easily refer to as 'similar persons in similar circumstances', as if there were no crucial question about what 'similar' means here.[14] At the same time, the argument's conclusion will be more likely to be misleading: it may no longer be a solution to *this* person's problem, so why should he let it weigh upon his deliberations of what he morally ought to do?

[14] Murdoch remarks: 'It is all very well to say that 'to copy a right action is to act rightly' (Hampshire, *Logic and Appreciation*), but what is the form which I am supposed to copy? It is a truism of recent philosophy that this operation of discerning the form is fairly easy […]' (Murdoch 2001: 29). Against 'similar persons in similar circumstances' arguments, see Gaita (2004: 67) and Rhees (1999: 50).

Even if we restrict the scope of moral arguments and understand them in an extremely particularistic way, as unique arguments for unique persons, there is a second sense in which thin arguments may fail to be personal, a sense that was already implicit in the previous section. An example from Gaita, already quoted in the previous chapter, will help to make it more explicit:

> If I am deliberating about which is the best route off the mountain and I fail to arrive at an answer, I can pass the problem over to my partner. It is only accidentally my problem. If I am deliberating about what morally to do, then I cannot pass my problem over to anyone else. It is non-accidentally and inescapably mine. […] If I seek your advice on which is the best route off the mountain, then the nature of what I do in asking for it, and what you do in giving it, is conditioned by the fact that I may hand the problem over to you completely (perhaps you are better at it, or perhaps I have lost my nerve), by the fact that I may consult manuals and by other such familiar facts. But if I must make a moral decision by Monday, I cannot come to you on Friday evening, plead that I have little time over the weekend to think about it, and ask you, a rational and informed agent and a professor of ethics to boot, to try to have a solution, or at least a range of options, no later than first thing on Monday morning. (Gaita 2004: 103)

Why are moral problems 'non-accidentally and inescapably mine'? It is not just because nobody happens to know me better than I know myself, because nobody is capable of taking my personal characteristics into account in a way and up to a degree that would be satisfactory. The problem here is not that others happen not to be able to do something for me, as if something were empirically implausible or impossible. It lies deeper and is more fundamental: I *ought* to think the problem through for myself. With moral problems comes the moral *task* to think about them. If Billy were to consult the smartest professor of ethics in the world and unthinkingly take whatever advice she gives, we would rightfully think that he has missed something about the kind of problem that he is dealing with. It is up to him to determine whether the reasons invoked by the professor (or the wise person, or whoever advises him) are 'enough'. Not because nobody else is capable of doing it for him, but simply because he ought to do it, and because it would be irresponsible and careless not to do it.[15] If, upon hearing that Billy was planning to consult a professor of ethics about his problem and to take her advice, whatever it may be, his wife was to feel scandalized or indignant, we would understand her. But if he doubts about whether the bird in his garden is a goldfinch or a goldcrest, and if in order to make really sure he would consult the smartest ornithologist in the world, he would, in normal circumstances, be justified in believing the ornithologist, and there would be no reason for his wife to be scandalized or indignant about it. Problems about moral advice and deference, the teachability of ethics and the possibility of action-guiding moral theory resurface here. It is this second sense of 'personal' that Wittgenstein, for example, emphasizes when he says that 'an ethical proposition is a personal act' (Wittgenstein 2003: 85), and it is the 'non-accidentally and inescapably mine' of moral problems that prompts him to say, in a conversation with members of the Vienna Circle:

[15] Compare Murdoch's description of someone's moral activity as 'peculiarly her own' (Murdoch 2001: 22).

> If I were told anything that was a *theory*, I would say, No, no! That does not interest me. Even if this theory were true, it would not interest me - it would not be the exact thing I was looking for.
>
> What is ethical cannot be taught. (Wittgenstein 1979: 116–117).

If moral problems are personal in the sense just outlined, if dealing with them has an irreducibly personal aspect, then moral arguments, even if put forward by experts, will often fail to give us reason to do whatever they conclude that we ought to do. Not because these arguments are likely to be invalid or because the conclusions are likely to be wrong, but *because* they conclude that we ought to do so and so. The problem is that the argument may come too close to a decision, like arguments that treat exploration problems as well-posed problems often do. It leaves no room for the personal task that Billy ought to accomplish. It ignores his responsibility in a way that both he and his wife may think unwelcome.[16]

A third sense in which moral arguments can fail to capture the personal element of moral problems concerns the person offering the argument rather than the person for whom it is meant (as in the first and second sense). This can be discerned in what Gaita says about moral advice. According to Gaita, 'we would not seek moral advice from someone whom we knew to be morally jaded', while 'being scientifically jaded [...] is of itself no bar to a scientist's authority to speak in his field' (Gaita 2004: 268).

> The difference has to do with what it is to have 'something to say' in each of these cases. We say of some people that they 'have something to say' on moral or spiritual matters, but we do not mean that they have information to impart or a theory to propound. We mean that they speak with an individual voice, but not because they know something that few people know. (Gaita 2004: 268)

In moral matters, Gaita continues, 'to have something to say is to be 'present' in what we say and to those to whom we are speaking', and 'what is said is not extractable from the manner of its disclosure' (Gaita 2004: 268). The conclusion of a moral argument 'must be someone's conclusion in a sense more substantial than is suggested by the fact that he feels compelled to write it at the end of a piece of reasoning on a blackboard' (Gaita 2004: 324).

Gaita points to a demand, when talking morally, to 'stand behind one's words' (Gaita 2004: 324) or, as Wittgenstein formulated it, 'to speak for oneself' (Waismann 1965: 16), 'to step forth as an individual and speak in the first person' (Wittgenstein 1979: 117). 'Thought which issues into a mere "blackboard conclusion", that is, a conclusion which seems inescapable when we write an argument on a blackboard, but which is a conclusion that we cannot seriously assert or even *wish* to assert, is thought that can have only the appearance of rigour' (Gaita 2004: 325). So what we expect from those offering moral arguments is not only that they take us into account as individuals and that they leave room for the individual moral task that comes with moral problems, but that they also speak to us with an individual voice. If they do

[16] The boundary between offering guidance and telling us what to do is blurred, and different persons confronted with the same argument may reasonably locate it on different sides of the boundary.

not, we will and should not let their arguments weigh on our considerations of what we morally ought to do. Not because their arguments are likely to be logically invalid or because their conclusions are likely to be wrong (although, if someone is morally jaded, the conclusion *is* likely to be wrong), since we are not concerned here with ways to determine whether a given argument is logically invalid or a conclusion wrong (which should not be taken to imply that it is not crucially important to determine this!), but because they will not be moral arguments in the thick sense. If Billy seeks advice from a professor of ethics, and if this professor is clearly morally jaded, then Billy will simply think that he has lost his time. What the professor says will not weigh with him, not speak to him, and so it cannot speak for or against a way of acting. And if it cannot do that, it does not deserve to be taken into consideration as a serious response to a moral problem.

8.6 An Absence of Morality

I have discussed four ways in which morality can be absent from moral arguments in the thin sense, ways that have been offered in a Wittgensteinian tradition of moral philosophizing (but not exclusively in this tradition).[17] Even if we have no independent reason to question the truth of the premises or the logical validity of the argument, a thin moral argument will fail to give us a reason to do whatever it claims that we ought to do if

1. The argument involves or leads to the denial of a moral certainty.
2. The argument displays no participant attitude.
3. The argument offers no deepened understanding of the problem.
4. The argument denies or ignores the personal character (in three different senses) of moral problems.

The emphasis has been on giving an overview of ways in which morality can be absent from moral arguments, on bringing them together within the framework of thin and thick moral arguments, on showing relations and highlighting distinctions between them. Undoubtedly, they each deserve a chapter- or book-length treatment that I cannot offer here, and I hope that the present chapter will encourage much-needed work in that direction.

In general, we can say that an absence of morality in thin arguments means that these arguments will fail to give us reason to do whatever they claim that we ought to do. The point is not so much that thin arguments are wrong. Rather, they are the wrong kind of arguments in response to people's moral problems. They pass the problem by or, put differently, do not adequately connect to the problem, and that is why we should not let them weigh upon our deliberations about what we morally ought to do. To bring that out, I have drawn upon a distinction between thin and

[17] Readers have remarked that, for example, Charles Taylor (1989) and Michael Walzer (1987, 1994) have made points similar to those developed in this chapter.

thick moral arguments. Thickness, of course, comes in degrees, and many moral arguments are more or less thick. That is, often the participant attitude is not completely absent, some deepened understanding is offered and the personal character of moral problems has more or less been taken into account. My main point can therefore be taken as a rather modest one, since it only involves claiming that arguments in which morality is *absent* in one or more of the four ways mentioned do not qualify as proper responses to real-life moral problems.[18]

One could remark that the thicker an argument is, the less general it is and the less application it has. The perfectly thick would be perfectly personal and the question is whether we can then still speak of an argument. If no degree of generality is allowed in moral arguments, then what is left for moral philosophers to do?[19] The answer to this objection is that I have not wished to claim that moral arguments should be, or aim to be, perfectly thick or perfectly personal. Some degree of generality *is* allowed in (and may even be necessary for) moral arguments, as long as their personal character has been taken into account. My point is not that 'the thicker a moral argument, the better', but only that a minimum of thickness is required if a moral argument is to give us reason to do whatever it claims that we ought to do. Moreover, thickness, although it has been my focus, is not the only criterion by which to judge moral arguments. The logical validity of the argument obviously plays a crucial role, too (albeit not the *only* crucial role, as I have wished to make clear). An argument may be sufficiently thick and logically invalid, or insufficiently thick and logically valid, and in both cases it will fail to give us reason to do whatever it claims that we ought to do. A good moral argument is both logically valid and reasonably thick. It appeals both on the argumentative and on the personal level.

The distinction between thin and thick moral arguments invites related distinctions that I have used, but not explicitly commented upon. Take, for example, the distinction between a moral issue as something that all moral arguments, thin and thick, deal with, and a moral problem as something that only thick moral arguments can properly deal with. According to this distinction, moral philosophers often deal with moral issues rather than with moral problems. A moral problem is here understood as the kind of problem people actually *having* the problem (people having to decide, being in the morally difficult situation) are dealing with, as a problem of life rather than a problem in philosophy. A moral issue (such as the permissibility of

[18] The absence of morality in one of these ways will often not be a matter of course: while Glock understands Singer's argument as leading to the denial of a moral certainty, some will deny that this is so. The distinction between thin and thick arguments is not sharp, but that does not make it useless. As we have seen in Chap. 4, Wittgenstein addresses the objection that 'a boundary [in our case, the boundary between thin and thick moral arguments] which is not sharply defined is not really a boundary at all' and that, consequently, we 'haven't accomplished anything at all' (Wittgenstein 2009a: §99). The vagueness of our boundary between thin and thick moral arguments is not to be denied, but it does not stand in the way of our boundary being a boundary. What it does stand in the way of is our being able to say without doubt in every conceivable case whether a moral argument is thick or merely thin. Even if this boundary is vague, that does not prevent us from showing that a certain argument falls on this or that side of it.

[19] I would like to thank an anonymous reviewer for pressing this objection.

abortion or euthanasia) is, in this sense of 'moral problem', not a moral problem for everyone (not, for instance, for most moral philosophers writing about it), and when it is a moral problem, it will be a problem about the permissibility of abortion in a particular case. These distinctions (although, of course, not always sharp) are often ignored, which leads to distorted ('over-philosophized') views of what it is to have a moral problem and what kind of response, answer or argument such problems ask for. Such distorted views reveal themselves in what Glock diagnoses as the uncanny willingness to follow thin moral arguments wherever they lead, or in what Gaita sees as a philosophers' 'ideal of intellectual purity which takes as its exemplar someone who would turn his life upside down after stumbling across an argument' (Gaita 2004: 317). The distinctions, reasons and reflections offered in this chapter may help to dispel the attractiveness of such ideals, for they show how to avoid confusion between moral issues and moral problems and the respective responses (thin and thick) that they require.

References

Blum, Lawrence A. 1994. *Moral Perception and Particularity*. Cambridge: Cambridge University Press.

Cavell, Stanley. 1999. *The Claim of Reason. Wittgenstein, Skepticism, Morality, and Tragedy*. Oxford: Oxford University Press.

de Waal, Frans. 2006. *Primates and Philosophers. How Morality Evolved*. Princeton: Princeton University Press.

Diamond, Cora. 2002. What if x isn't the number of sheep?' Wittgenstein and Thought-Experiments in Ethics. *Philosophical Papers* 31: 227–250.

Gaita, Raimond. 2004. *Good and Evil. An Absolute Conception*. 2nd ed. London: Routledge.

Geertz, Clifford. 1973. *The Interpretation of Cultures. Selected Essays*. New York: Basic Books.

Glock, Hans-Johann. 2008. *What Is Analytic Philosophy?* Cambridge: Cambridge University Press.

Hacker, P.M.S. 2009. Philosophy. A Contribution Not to Human Knowledge but to Human Understanding. In *The Nature of Philosophy*, Royal Institute of Philosophy Supplement 65, ed. Anthony O'Hear, 129–153. Cambridge: Cambridge University Press.

Lovibond, Sabina. 1983. *Realism and Imagination in Ethics*. Oxford: Blackwell.

Murdoch, Iris. 2001. *The Sovereignty of Good*. London: Routledge.

Pleasants, Nigel. 2008. Wittgenstein, Ethics and Basic Moral Certainty. *Inquiry* 51: 241–267.

———. 2015. If Killing Isn't Wrong, Then Nothing Is. A Naturalistic Defence of Basic Moral Certainty. *Ethical Perspectives* 22: 197–215.

Rhees, Rush. 1965. Some Developments in Wittgenstein's View of Ethics. *The Philosophical Review* 74: 17–26.

———. 1999. *Moral Questions*. Basingstoke: MacMillan Press.

Rummens, Stefan. 2013. On the Possibility of a Wittgensteinian Account of Moral Certainty. *The Philosophical Forum* 44: 125–147.

Ryle, Gilbert. 2009. The Thinking of Thoughts. What Is 'Le Penseur' Doing? In *Collected Papers. Volume 2. Collected essays 1929–1968*, 494–510. London: Routledge.

Singer, Peter. 2009. Speciesism and Moral Status. *Metaphilosophy* 40: 567–581.

Strawson, P.F. 2009. Freedom and Resentment. In *Free Will*, ed. Derk Pereboom, 148–171. Indianapolis: Hackett.

Taylor, Charles. 1989. *Sources of the Self. The Making of the Modern Identity*. Cambridge, MA: Harvard University Press.
Waismann, Friedrich. 1965. Notes on Talks with Wittgenstein. *The Philosophical Review* 74: 12–16.
Walzer, Michael. 1987. *Interpretation and Social Criticism*. Cambridge, MA: Harvard University Press.
———. 1994. *Thick and Thin. Moral Argument at Home and Abroad*. Notre Dame: University of Notre Dame Press.
Williams, Bernard. 1985. *Ethics and the Limits of Philosophy*. Cambridge, MA: Harvard University Press.
Wittgenstein, Ludwig. 1974. *On Certainty*. Oxford: Blackwell.
———. 1979. In *Wittgenstein and the Vienna Circle*, ed. Brian McGuinness. Oxford: Blackwell.
———. 2003. In *Public and Private Occasions*, ed. James C. Klagge and Alfred Nordmann. Lanham: Rowman & Littlefield.
———. 2009a. *Philosophical Investigations*. 4th ed., Trans. G.E.M. Anscombe, P.M.S Hacker, and Joachim Schulte. Oxford: Wiley-Blackwell.
———. 2009b. Philosophy of Psychology – A Fragment. In *Philosophical Investigations*, 182–244. 4th ed., Trans. G.E.M. Anscombe, P.M.S Hacker, and Joachim Schulte. Oxford: Wiley-Blackwell.
———. 2014. A Lecture on Ethics. In *Lecture on Ethics*, ed. Edoardo Zamuner, Ermelinda Valentina, Di Lascio, and D.K. Levy, 42–51. Oxford: Wiley-Blackwell.

Chapter 9
Conclusion

I discuss three ways in which the topic of this book can be developed in future research and, for each of these ways, provide examples of how such research could proceed and in which contemporary debates it could make a difference. I conclude that contemporary mainstream moral philosophy has much to gain by paying (more) attention to the later Wittgenstein's philosophical thought. I also provide some suggestions as to why Wittgenstein's later philosophy and the Wittgensteinian tradition are almost absent in contemporary moral philosophy.

With this book I hope to have contributed to (1) the understanding of Wittgenstein's later philosophy, (2) the understanding of his ethical views in the light of his later philosophical thought and (3) the understanding of several ways in which his later philosophical thought can be applied to several problems in contemporary moral philosophy.

The chapters of this book do not correspond to the steps of an argument in which each step logically follows from the previous step and logically leads to the next. The same three goals could have been achieved in an entirely different way, with seven different chapters covering seven different debates. It has not been my aim to prove, by way of a general argument, the correctness of the thesis that Wittgenstein's later philosophy is helpful in dealing with problems in moral philosophy, and I would not know how to go about proving that. Rather, my aim has been to provide worked-out examples of *how* Wittgenstein's later thought can be applied to problems in moral philosophy.

It is not a coincidence, in that regard, that the book is a collection of articles. Wittgenstein writes in the Preface of the *Investigations* that it seems essential that in a book 'the thoughts should proceed from one subject to another in a natural, smooth sequence', while 'the very nature of the investigation […] compels us to travel criss-cross in every direction over a wide field of thought' (Wittgenstein 2009: 4). He describes the remarks in his book as 'a number of sketches of landscapes' (Wittgenstein 2009: 4) whose aim is 'to give the viewer an idea of the landscape. So this book is really just an album' (Wittgenstein 2009: 5). I believe that 'really just an album' is an apt description of this book. Although I hope that within the chapters

B. De Mesel, *The Later Wittgenstein and Moral Philosophy*, Nordic Wittgenstein Studies 4, https://doi.org/10.1007/978-3-319-97619-8_9

the thoughts proceed in a more or less natural sequence, the relation between the different chapters, especially between the different chapters within a single part of the book, is much more like the relation between sketches of a landscape than the relation between the steps of a formal argument. As Wittgenstein says: ‘I jump about all round the topic; that is the only way of thinking that is natural to me’ (Wittgenstein 1998: 33).

Precisely because of its open structure, it should not be difficult to see that it would be inappropriate to end this work with an overall conclusion that closes things off. Marie McGinn opens her book on the *Investigations* with a quotation from Gustave Flaubert: ‘Stupidity consists in wanting to reach conclusions. [...] What mind worthy of the name, beginning with Homer, ever reached a conclusion?’ (McGinn 1997: 5). This quotation is fitting for a book on the later Wittgenstein. The nature of his investigations, best captured in sketches of landscapes, does not support grand conclusions. Such an overall conclusion would be *the* ultimate drawing of (a part of) the landscape, but there is no such thing.

That does not mean that we are left with nothing to say. First, we can at least say that the conclusion of this book is the sum of the conclusions of its chapters. These are, roughly, that (1) there are good reasons to think that a Wittgensteinian contribution to moral philosophy may not only lie in an elaboration of Wittgenstein’s ethical views, but also, and maybe primarily, in the application of his later methods to problems in moral philosophy. (2) Concerning these later methods, the comparison of philosophical methods to therapies has had a misleading effect because properties of therapies have been illegitimately projected onto Wittgenstein’s methods. (3) It is possible and useful for moral philosophy to provide surveyable representations of moral vocabulary. (4) Wittgenstein does not exclude the possibility of doing meta-ethics and of there being a particular subject matter of moral philosophy. (5) There is room for objectivity in Wittgenstein’s ethical views. (6) Wittgenstein’s remarks on aspect-seeing and seeing emotion provide a useful object of comparison for thinking about moral perception. (7) Questions for moral advice of the form ‘Should I do x or y?’ are often importantly different from non-moral questions of that form. (8) Arguments provided by moral philosophers often fail to give us a reason to do whatever they claim that we ought to do, even if we see no independent reason to question the truth of the premises or the logical validity of the argument.

Second, these intermediate conclusions support the more general conclusion that an application of Wittgenstein’s later methods to certain problems in moral philosophy may help to better understand and even dissolve these problems. Such a conclusion, however, is not meant to close discussion on the topic. Instead, I hope that it will be considered as an invitation to provide more worked-out examples and that it will encourage philosophical work in several directions.

First, although I have been focusing on the *Philosophical Investigations,* I have not been able to do full justice to the richness of that book. Several aspects of it that I have barely touched upon may well be applicable to problems in moral philosophy. Take, for example, the much-debated rule-following considerations and the dispute between generalists and particularists in moral philosophy about the question of whether morality does or does not need rules or principles. As we have seen,

Wittgenstein emphasizes again and again that we use words 'in an enormous number of different and differently related ways' (Wittgenstein 2009: §47). He provides the example of the word 'composite' and adds: 'To the *philosophical* question "Is the visual image of this tree composite, and what are its constituent parts?" the correct answer is: "That depends on what you understand by 'composite'"' (Wittgenstein 2009: §47). Similarly, the words 'rule' and 'principle' are used in an enormous number of different and differently related ways, as the rule-following considerations bring out, and whether morality is principled or needs rules will depend on what one understands by 'principle' or 'rule'. If generalists understand something different by 'principle' or 'rule' than particularists do, it should not come as a surprise that they draw different conclusions as to whether morality is principled. These different conclusions will then, however, no longer be in conflict with each other, as they are commonly thought to be.

Although I cannot provide more than a rough indication here, I indeed believe that the conflict between particularists and generalists is not a conflict about the nature of morality, but rather a conflict about the meaning of 'principle' and 'rule'. According to Jonathan Dancy, the most prominent particularist,

> The principle that it is wrong to lie, for instance, presumably claims that mendacity is always a wrong-making feature wherever it occurs (that is, it always makes the same negative contribution, though it often does not succeed in making the action wrong overall). (Dancy 2004: 76)

Thus, Dancy thinks that the moral principle 'It is wrong to lie' necessarily involves the claim that lying always makes an action morally worse. Generalists such as Sean McKeever and Michael Ridge deny that (McKeever and Ridge 2006: Chapters 1 and 2) and work with another, broader conception of 'principle'. Both parties seem to be asking the same question ('Is morality principled?') and give different answers to it, but I suspect that in fact they are asking different questions and are talking past one another. I fully agree with Maike Albertzart that 'there is no point in discussing whether or not ethics should be principled without having first reached an agreement on our understanding of the nature and role of moral principles' (Albertzart 2011: 58). Here, a Wittgensteinian approach could help. What is needed is a surveyable representation, a description of the use of 'principle' and 'rule', of the part that principles play in our life (Wittgenstein 2009: §156), of 'the game with these words, their use in linguistic intercourse that is carried on by their means' (Wittgenstein 2009: §182). It seems likely that the rule-following considerations (Wittgenstein 2009: §§185–242) will provide helpful material with regard to our use of the word 'rule'.

The possibility of using the rule-following considerations in order to dissolve the debate between particularists and generalists is only one example of the manifold possibilities for applying Wittgenstein's later thought to problems in moral philosophy that I have not discussed in detail. It is meant to show that these possibilities are anything but exhausted. More aspects of the *Investigations* can be applied to, and enhance our understanding of, more of Wittgenstein's ethical views and more debates in moral philosophy.

Second, I have used material from the whole of Wittgenstein's later work, but the emphasis was on the methods that Wittgenstein uses and explicitly discusses in the *Philosophical Investigations*. However, Wittgenstein's later work is much richer than the *Investigations* alone, and I think that my references to these works, though not as numerous as those to the *Investigations*, provide at least some preliminary support for that claim.

In the context of other later works and the possible application of methods to problems in moral philosophy, I should especially mention *On Certainty*. In *On Certainty*, Wittgenstein only discusses (what we could, roughly speaking, call) empirical certainties, but it may well be that, as Judith Lichtenberg says, 'his explorations of the concepts of knowledge, certainty, belief, and justification in the context of non-moral judgements help illuminate the realm of the moral as well' (Lichtenberg 1994: 184). Several commentators have investigated the possibility of moral certainties on the model of Wittgenstein's empirical certainties (Brice 2013; Harré 2009, 2010; Hermann 2015; Kober 1997; Lichtenberg 1994; Pleasants 2008, 2009, 2015; Rummens 2013). I believe that the idea of moral certainties is an interesting one, although it has its problems. Some of these have been mentioned in Chap. 8, Sect. 8.2, but it seems evident that more work on what I called the 'problem of criticism', and on the compatibility of the idea of moral certainties with the idea of moral revolutions, is needed.

I have mentioned two directions in which the topic of this book can (and should, in my view) be further developed. First, other aspects of the *Investigations* can be applied to other problems in moral philosophy. Second, Wittgenstein's later work, and especially *On Certainty*, has much more to offer to moral philosophy than the *Investigations* alone. My third proposal goes beyond Wittgenstein. Many philosophers have been doing work in moral philosophy that bears, to a greater or lesser extent, the mark of Wittgenstein's influence. I am thinking here of such authors as Stanley Cavell, Cora Diamond, Raimond Gaita, Sabina Lovibond, John McDowell, Iris Murdoch, D.Z. Phillips, Rush Rhees, David Wiggins and Peter Winch. I believe that their ideas, views and ways of approaching moral problems are underrepresented in contemporary mainstream moral philosophy, to which they could make an interesting contribution (see, for suggestions, Chap. 8). That is the third direction in which the topic of this book can be further developed.

In general, the Wittgensteinian tradition raises (or revives) crucial questions with regard to the relation between moral theories and people's actual moral problems. One perennial debate in moral philosophy to which Wittgensteinian ideas and methods could make a valuable contribution concerns the debate about the possibility of action-guiding moral theory. Wittgensteinian work on that topic has already been carried out (Chappell 2009; Crary 2007; Hämäläinen 2010; Phillips 1992; Rhees 1999), but it should be clear that, as long as action-guiding moral theories are being put forward, there is an ongoing philosophical task to investigate their possible relations to people's actual moral lives and problems. Another debate in which the Wittgensteinian tradition has something to say is the debate about moralism (Cavell 1999: 253–254; Crary 2007: Chapter 6; Diamond 1997; Taylor 2012). The danger of moralism pervades much of moral philosophy, and it seems important for moral philosophers to have an adequate understanding of that danger. The third and last debate (or group of debates) that I will mention here concerns the issues of moral advice, moral testimony and moral expertise. These are much-discussed topics both in contemporary mainstream moral philosophy (for helpful references, see Hills 2013; for recent contributions, see Davia and Palmira 2015 and Enoch 2014) and in the Wittgensteinian tradition (Gaita 2004; Phillips 1964), but the points made in the Wittgensteinian tradition have barely found their way into the contemporary debate, while I am confident that they could enrich it.

It should by now be clear that I firmly believe that contemporary mainstream moral philosophy has much to gain by paying (more) attention to the later Wittgenstein's philosophical methods and the Wittgensteinian tradition. I even hope that, in the light of the arguments for doing so that I have given, it may start to sound strange that both Wittgenstein's later methods and the Wittgensteinian tradition are almost absent there. However, the relative absence of later Wittgensteinian thought is not only discernible in moral philosophy. According to John Cottingham, 'Wittgenstein's influence has recently suffered something of an eclipse in the Anglophone philosophical world' (Cottingham 2009: 203), and P.M.S. Hacker says that 'his [Wittgenstein's] ideas are on the whole neglected by leading figures in contemporary philosophy' (Hacker 2013: xvii). The question is why. Cottingham suggests three possible answers.

> This may well be a natural 'rebound' reaction against the climate of that substantial chunk of the twentieth century when much of philosophy was dominated by his [Wittgenstein's] approach to the subject. It may also be a result of a certain cautious, academic tidy-mindedness, which is wary of work that is sweeping enough to resist neat dissection within the burgeoning technical specialisms of current 'mainstream' philosophy. Or, thirdly, it may be due to the rise of a scientistic vision of philosophy […]. (Cottingham 2009: 203)

Hacker too regards the growth of scientism and the acceptance of continuity between philosophy and science as factors that explain why approaches in the spirit of the later Wittgenstein are not very popular in contemporary mainstream philosophy. He adds four more reasons for their relative neglect.

> […] (a) the emergence, in the wake of Quine's work, of new kinds of philosophy of language [and these, in turn, 'prepared the way for the revival of metaphysics, the impetus for which was provided by Lewis and Kripke' (Hacker 2013: xx)]; (b) the simultaneous development of the new theoretical linguistics at the hands of Chomsky and his followers; (c) the computer revolution, which inclined philosophers, scientists and psychologists to construe the operations of the mind and the workings of the brain on the model of computers; (d) advances in neurophysiological psychology, which, at a highly speculative level, seemed to invite an attempted synthesis of theoretical linguistics, philosophy of language, artificial intelligence and cybernetics, psychology and philosophy of mind – a trend culminating in the invention of a new subject, representational 'cognitive science'. (Hacker 1996: 268)

According to Hacker, these four factors, 'in different ways, fostered complementary features of thought which can be said to be "in the spirit of the *Tractatus*", even though often vulgarized' (Hacker 1996: 268).

In general, Hacker's complaint is that contemporary philosophy has in many ways returned to views and methods that Wittgenstein held and practiced in the

Tractatus but showed to be gravely mistaken in the *Investigations* and other later works. Historically speaking, aspects of Tractarian thought were brought to the United States by ex-members and affiliates of the Vienna Circle such as Rudolf Carnap. Hacker claims that the impact of these emigrés 'gave American philosophy its characteristic profile for the rest of the twentieth century' (Hacker 1996: 183). By the mid-1970s, and partly due to economic reasons, 'the centre of Anglophone philosophy became the United States. There were more philosophers teaching there than the sum of philosophers in the whole history of the subject' (Hacker 1996: 265). Consequently, the Tractarian spirit of American philosophy came to be dominant in the Anglophone philosophical world. Aspects of Wittgenstein' later thought that were not compatible with this spirit were pushed to the margins of philosophical practice. It is clear for Hacker that 'the rejection of Wittgenstein's philosophy and methodology has not been the result of the refutation of his ideas and the proven inadequacy of his methods. Indeed, it has not even rested on comprehension of his ideas' (Hacker 2013: xx).

I cannot here evaluate in depth the reasons that Cottingham and Hacker mention for the relative absence of Wittgenstein and the Wittgensteinian tradition in contemporary mainstream philosophy, but I am inclined to basically agree with their diagnoses. I would like to add one more element that Hacker also emphasizes. Not only those who have more or less ignored Wittgenstein's later thought are responsible for its lack of impact nowadays. Many Wittgensteinians too have made mistakes in that respect. They have been involved in endless exegetical discussions about sorts of nonsense in the *Tractatus* and allowed themselves 'to become distracted from the serious task of trying to interpret his philosophy of language, his philosophy of logic and mathematics, and his philosophy of psychology' (Hacker 2013: xvii). 'This debate on what Geach called "Ludwig's self-mate"', Hacker claims, 'has not fertilized other philosophical investigations, as did previous debates on family resemblance, language-games, rule-following, criteria, and private languages' (Hacker 2013: xviii). As a result, 'few attempt to apply his [Wittgenstein's] methods to new domains in philosophy or in conceptual criticism of the natural sciences, the sciences of the mind and brain, and the social sciences for which criticisms he gave both a rationale and a warrant' (Hacker 2013: xvii).

Hacker is not the only one to accuse Wittgensteinians of isolating themselves, of locking themselves up in their own world, of endlessly debating each other's work on Wittgenstein while remaining aloof from debates that concern the larger philosophical community. These accusations are, in my view, too often justified. To be clear, I am not saying that exegesis is not needed. On the contrary, it *is* still needed, and I have devoted a significant part of this book to it. However, it is not enough, and I wholeheartedly agree with what Peter Winch writes in his introduction to *Trying to Make Sense* (1987). Although some of his essays in that book are explicitly exegetical, Winch describes his attitude to Wittgenstein's work as an attitude 'of gratitude for the help it has given me in seeing what are the important questions, and what kinds of questions they are, rather than that of an aspiring exegete' (Winch 1987: 1). I believe that an attitude such as Winch's has to be fostered. One can hardly expect to convince philosophers with little knowledge of or interest in

Wittgenstein of the philosophical value of his later work by showing them a piece of exegesis, especially not when it has no bearing on prominent philosophical debates. What could convince them is a fresh approach to problems they are working on, an approach that enhances their understanding of the questions they are dealing with and the ways in which these can be profitably elucidated. I hope that this book goes at least some way in showing how such an approach could work with regard to several problems in moral philosophy, a domain (but not the only one) where Wittgensteinian methods are much needed. Let me close by quoting Wittgenstein: 'The works of the great masters are stars which rise and set around us. So the time will come again for every great work that is now in the descendent' (Wittgenstein 1998: 23).

References

Albertzart, Maike. 2011. Missing the Target. Jonathan Dancy's Conception of a Principled Ethics. *The Journal of Value Inquiry* 45: 49–58.

Brice, Robert. 2013. Mistakes and Mental Disturbances. Pleasants, Wittgenstein, and Basic Moral Certainty. *Philosophia. Philosophical Quarterly of Israel* 41: 477–487.

Cavell, Stanley. 1999. *The Claim of Reason. Wittgenstein, Skepticism, Morality, and Tragedy*. Oxford: Oxford University Press.

Chappell, Timothy. 2009. Ethics Beyond Moral Theory. *Philosophical Investigations* 32: 206–243.

Cottingham, John. 2009. The Lessons of Life. Wittgenstein, Religion and Analytic Philosophy. In *Wittgenstein and Analytic Philosophy. Essays for P.M.S. Hacker*, ed. Hans-Johann Glock and John Hyman, 203–227. Oxford: Oxford University Press.

Crary, Alice. 2007. *Beyond Moral Judgment*. Cambridge, MA: Harvard University Press.

Dancy, Jonathan. 2004. *Ethics Without Principles*. Oxford: Clarendon.

Davia, Cory, and Michele Palmira. 2015. Moral Deference and Deference to an Epistemic Peer. *The Philosophical Quarterly* 65: 605–625.

Diamond, Cora. 1997. Moral Differences and Distances. Some Questions. In *Commonality and Particularity in Ethics*, ed. Lili Alanen, Sara Heinämaa, and Thomas Wallgren, 197–234. Basingstoke: Macmillan.

Enoch, David. 2014. A Defense of Moral Deference. *The Journal of Philosophy* 111: 229–258.

Gaita, Raimond. 2004. *Good and Evil. An Absolute Conception*. 2nd ed. London: Routledge.

Hacker, P.M.S. 1996. *Wittgenstein's Place in Twentieth Century Analytic Philosophy*. Oxford: Blackwell.

———. 2013. *Wittgenstein. Comparisons and Context*. Oxford: Oxford University Press.

Hämäläinen, Nora. 2010. Is Moral Theory Harmful in Practice? Relocating Anti-Theory in Contemporary Ethics. *Ethical Theory and Moral Practice* 12: 539–553.

Harré, Rom. 2009. Wittgenstein's Therapies. From Rules to Hinges. *New Ideas in Psychology* 27: 118–132.

———. 2010. Are There Moral Hinges? *Praxis. Revista de Psicologia* 18: 11–27.

Hermann, Julia. 2015. *On Moral Certainty, Justification and Practice. A Wittgensteinian Perspective*. Basingstoke: Palgrave Macmillan.

Hills, Alison. 2013. Moral Testimony. *Philosophy Compass* 8: 552–559.

Kober, Michael. 1997. On Epistemic and Moral Certainty. A Wittgensteinian Approach. *International Journal of Philosophical Studies* 5: 365–381.

Lichtenberg, Judith. 1994. Moral Certainty. *Philosophy* 69: 181–204.

McGinn, Marie. 1997. *Routledge Philosophy Guidebook to Wittgenstein and the Philosophical Investigations*. London: Routledge.
McKeever, Sean, and Michael Ridge. 2006. *Principled Ethics. Generalism as a Regulative Ideal*. Oxford: Clarendon.
Phillips, D.Z. 1964. The Possibilities of Moral Advice. *Analysis* 25: 37–41.
———. 1992. *Interventions in Ethics*. Basingstoke: Macmillan.
Pleasants, Nigel. 2008. Wittgenstein, Ethics and Basic Moral Certainty. *Inquiry* 51: 241–267.
———. 2009. Wittgenstein and Basic Moral Certainty. *Philosophia. Philosophical Quarterly of Israel* 37: 669–679.
———. 2015. If Killing Isn't Wrong, Then Nothing Is. A Naturalistic Defence of Basic Moral Certainty. *Ethical. Perspectives* 22: 197–215.
Rhees, Rush. 1999. *Moral Questions*. Basingstoke: MacMillan Press.
Rummens, Stefan. 2013. On the Possibility of a Wittgensteinian Account of Moral Certainty. *The Philosophical Forum* 44: 125–147.
Taylor, Craig. 2012. *Moralism. A Study of a Vice*. Durham: Acumen.
Winch, Peter. 1987. *Trying to Make Sense*. Oxford: Blackwell.
Wittgenstein, Ludwig. 1998. *Culture and Value*. Rev. 2nd ed. Oxford: Blackwell.
———. 2009. *Philosophical Investigations*. 4th ed., Trans. G.E.M. Anscombe, P.M.S Hacker, and Joachim Schulte. Oxford: Wiley-Blackwell.

Index

A
Absolute value, 61
Advice, 22, 23, 94, 98, 132, 133, 135, 137, 138, 142–144, 147–149, 167–170, 176, 179
Aesthetics/aesthetical, 4, 59–61, 85, 87, 164
Albertzart, M, 177
Analogy, 12, 15, 20, 21, 34, 39, 45, 54, 56, 79, 81, 95, 100, 101, 108, 113–118, 124, 125, 127–129, 156, 157
Andronico, M., 85, 86
Anscombe, G.E.M., 51, 142, 162
Arendt, H., 136, 137
Argument, 175, 176, 179
Aspect perception/aspect-seeing, 4, 20, 22, 114, 116, 118–122, 126–128, 176
Attitude, 23, 38, 77, 85, 93, 94, 96, 97, 100, 102, 138, 144, 154, 160–163, 165, 166, 170, 171, 180
Audi, R., 115, 118, 121
Authority, 3, 138, 140, 169
Autonomy/autonomous, 140, 142, 144, 146

B
Baker, G.P., 7, 8, 14–16, 18, 34, 40, 51–59, 98, 100, 121, 124
Baz, A., 114, 145
Blackburn, S., 114, 118, 122, 123, 125, 127, 128
Blind, 122–124
Blum, L., 161
Bouveresse, J., 39, 41
Brandhorst, M., 19, 91
Brand, R., 38, 43

C
Cavell, S., 1, 23, 86, 97, 154, 160, 163–166, 178
Certainty, 4, 23, 154–161, 165, 170, 171, 178
Chappell, T., 128, 178
Churchill, J., 44
Cognitivism, 64, 66, 101, 128
Colour, 20, 21, 51, 52, 54, 56, 58, 108, 113–129
Comparison, 13–16, 21, 23, 31–46, 52, 56–58, 60, 64, 118, 135, 176
Conant, C., 34, 43, 44, 72, 75, 77, 78, 84–86
Concept/conceptual, 4, 12, 13, 15, 16, 18, 21, 22, 32, 34, 36–39, 42, 44, 49–59, 63, 64, 77–80, 83, 87, 94, 97, 99, 102, 104, 107, 116, 120–122, 147, 156, 157, 159, 164, 166, 178, 180
Consequentialism/consequentialist, 162
Conventionalism/conventionalist, 158
Cottingham, J., 179, 180
Crary, A., 1, 5, 97, 99, 147, 148, 178

D
Dancy, J., 177
D'Arms, J., 115, 118, 126
Deflationism/deflationary, 94, 98–104
de Waal, F., 158

B. De Mesel, *The Later Wittgenstein and Moral Philosophy*, Nordic Wittgenstein Studies 4, https://doi.org/10.1007/978-3-319-97619-8

Diamond, C., 1, 17, 20, 23, 50, 62, 65, 72–78, 80–84, 89, 97, 99, 100, 140, 148, 154, 161–163, 165, 178
Dilemma, 64–66, 82, 87, 89, 109, 148, 149, 161, 162, 165
Disorder, 15, 32–38, 40, 42, 44
Dogmatism/dogmatic, 7, 8, 20, 161
Doubt, 4, 6, 10, 19, 32, 65, 81, 87, 99, 104, 126, 133, 136, 155–157, 159, 163–166, 168, 171

E
Education, 120–122, 127
Emotion, 17, 21, 64–66, 113, 116, 117, 176
Empirical, 24, 41, 65, 66, 156–159, 165, 178
Ethics, 2, 51, 71, 91, 119, 134, 176
See also Moral philosophy
Evidence, 2, 24, 32, 35, 45, 51, 65, 116, 121, 157, 158
Expertise, 134, 167, 179

F
Fact, 2, 4, 6, 10, 12, 19, 21, 34, 36–39, 43, 44, 55, 58–60, 67, 72, 80, 83, 91–97, 99, 100, 102–106, 109, 114, 119, 120, 122, 125, 127, 128, 131, 133, 138, 140–142, 146–149, 154, 158–160, 162, 165, 168, 169, 177
Falsity, 19, 91, 99
Family resemblance, 4, 12, 13, 42, 180
Fischer, E., 34, 38, 41, 55
Fisher, A., 96, 99, 119, 126
Flaubert, G., 176
Form of life, 87, 101, 159
Freud, S., 39, 41

G
Gaita, R., 1, 23, 97, 131, 132, 138–140, 142, 143, 154, 158, 159, 162, 165–169, 172, 178, 179
Galilei, 157, 158
Gandhi, M., 158
General facts of nature, 4
Generalism/generalist, 167, 176, 177
Generality, 43, 171
Gentner, D., 15
Gert, B., 149
Gill, M., 64, 66
Gilligan, C., 161
Glock, H.-J., 9, 10, 23, 24, 39, 53, 56, 57, 82, 121, 153–158, 160, 171, 172
Goldie, P., 114, 121
Grammar/grammatical, 37, 39, 50–54, 56–58, 60, 61, 64, 67, 74, 80, 82, 89

H
Hacker, P.M.S., 7, 8, 14, 16, 18, 34, 39, 40, 51–59, 62, 64, 98, 100, 119, 157, 158, 179, 180
Hanfling, O., 19, 105
Hardy, G.H., 95
Health, 15, 31–33, 43–45
Hertzberg, L., 72, 74–76, 84, 85, 142
Holtzman, S., 97, 98, 100
Horwich, P., 43
Hutchinson, P., 15, 34, 35, 39

I
Illness, 15, 31–38
Impersonal, 24, 80, 93, 96, 102, 137, 166, 167

J
Jacobson, D., 115, 118, 126
Jacquette, D., 36, 42, 44
Johnston, P., 1, 50, 62, 80, 99, 143
Jolley, K.D., 121, 122
Judgment, 17, 21, 22, 65, 66, 89, 91, 94, 97, 98, 102, 103, 105, 116, 128, 134, 137, 140, 146, 147, 149, 156, 160
Justification, 4, 21, 98, 119, 156, 178

K
Kirchin, S., 96, 99, 119, 126
Know/knowledge, 5, 10, 21, 23, 24, 35, 39, 41, 53, 54, 56, 57, 59–61, 65, 67, 76, 81, 84, 99, 100, 102, 104, 108, 115, 123, 124, 126, 129, 133–135, 137–139, 142, 144, 145, 148, 158, 159, 163, 167–169, 175, 178, 180
Kuusela, O., 1, 7, 8, 14, 37, 42, 52, 54, 55, 57, 67, 89

L
Language-game, 2, 4, 5, 13, 19, 33, 52, 54–57, 89, 94, 98, 99, 145
Leibowitz, U., 143, 147–149
Leich, C., 97, 98, 100
Lichtenberg, J., 178

Logic, logical, 1, 5, 7, 13, 23, 33, 58, 62, 80, 95, 100, 101, 105, 153, 154, 161, 167, 170, 171, 176, 180
Lovibond, S., 1, 23, 72, 73, 87, 89, 97, 100, 157–159, 178

M
Majetschak, S., 39, 40
Mandela, N., 158
Mathematics/mathematical, 1, 2, 5, 13, 17, 19, 36, 37, 42, 72, 81, 82, 94, 95, 97, 100, 101, 105–107, 162, 180
McDowell, J., 97, 114–118, 120, 126, 128, 178
McGinn, M., 14, 176
McKeever, S., 113, 177
Meaning, 5, 7, 13, 17, 19, 23, 72, 73, 78, 82, 103, 131, 133, 137, 139–141, 143, 145, 146, 148, 154, 164, 165, 177
Meta-ethics, 19, 20, 63, 64, 71–89, 176
Metaphor, 60
Metaphysics/metaphysical, 7, 12, 19, 33, 39, 86, 94, 95, 99, 100, 104, 179
Method, 1, 4–11, 13–18, 21, 24, 31–46, 49, 50, 53, 55–59, 62, 63, 65, 66, 88, 154, 176, 178–181
Modesty, 22, 143, 149
Monk, R., 6, 36, 45, 58, 62
Moore, G.E., 57, 59, 76, 78, 79, 83, 149
Moral advice, 22, 137, 138, 143, 144, 147–149, 167–169, 176, 179
Moral argument, 23, 24
Moral blindness, 4, 122–124
Moral certainty, 4, 23, 154–161, 165, 170, 171
Moral deference, 134, 167
Moral dilemma, 89, 148, 149
Moral expertise, 134, 167, 179
Moral judgment, 17, 22, 64–66, 89, 97, 134, 137, 140, 146, 147, 149, 156
Moral modesty, 22, 143, 149
Moral perception, 20–22, 113–116, 118, 120–122, 125–129, 176
Moral philosophy, 1, 49, 72, 143, 153, 175
See also Ethics
Moral problem, 2, 3, 23, 24, 63, 64, 132, 153, 154, 160, 161, 163, 165–172, 178
Moral property, 72, 75
Moral revolution, 158, 159, 178
Moral testimony, 179
Moral theory, 149, 162, 167, 168, 178
Moral use, 19, 50, 51, 62, 63, 65, 66, 74, 83, 84, 88
Moral value, 20–22, 113
Moral vocabulary, 49, 72–76, 78, 83, 176
Moral word, 50, 59, 61, 72–76
Moralism, 178
Morality, 23, 49, 75, 106, 147, 149, 176, 177
Moyal-Sharrock, D., 6, 20
Mulhall, S., 72, 75, 77–81, 83, 85, 87, 88
Murdoch, I., 23, 97, 164, 167, 168, 178

N
Neutral, neutrality, 80, 88, 93, 96, 102, 161
Non-cognitivism, 64, 66
Nonsense/nonsensical, 43, 51, 54, 58, 60–62, 76, 180
Normativity/normative, 3, 23, 63, 64, 80, 81, 88, 101, 153

O
Oberdiek, H., 64
Object of comparison, 21, 45, 55, 59, 118, 176
Objective/objectivity, 19, 20, 37, 91–109, 116, 117, 160, 166, 176

P
Pain, 12, 13, 42, 66, 97, 104, 116, 117
Paradigm, 22, 52, 117, 134
Particularism/particularist, 167, 176, 177
Patient, 15, 31–33, 36, 45, 155
Perception, 20, 113, 114, 176
Personal, 2–4, 23, 24, 34, 38, 62, 67, 80, 93, 96, 97, 102, 103, 107, 108, 137, 138, 147, 154, 166–171
Peterman, J., 37, 39
Phillips, D.Z., 143, 178, 179
Philosophical methods, 1, 4, 9–11, 14–17, 31–46, 50, 55, 56, 58, 176, 179
Philosophy, 1–5, 9, 14, 17–20, 31–35, 37–46, 49–67, 71–89, 97, 100, 101, 105, 128, 129, 143, 147, 153, 154, 157, 159, 167, 171, 175–181
Philosophy of language, 1, 5, 13, 77–81, 179, 180
Philosophy of mind, 1, 13, 77–80, 97, 179
Physical, 15, 42, 91, 93, 101, 105–107, 157
Physics, 93, 95, 96, 100, 104
Picture, 5, 12, 38, 46, 51, 56, 65, 91, 92, 102, 104, 106, 107, 127
Plant, B., 36, 74
Platonism/Platonist, 91, 92, 95, 100–104, 106, 107
Pleasants, N., 123, 156, 178
Plutchik, R., 127

Practice, 8, 12, 13, 24, 41, 55, 97, 149, 155, 180
Private language argument, 4
Psychoanalysis, 39–41
Psychology/psychological, 13, 15, 16, 31, 33, 36, 39–45, 52, 106, 179, 180
Subjective/subjectivity, 97–100, 102, 103, 115–117, 166
Subject matter, 17, 18, 58, 64, 67, 71–89, 99, 176
Surveyable representation, 13, 16, 41, 49–67, 176, 177

R
Railton, P., 114, 125
Read, R., 6, 34, 35, 39
Realism/realist, 72, 100–102, 104, 117, 118, 128
Reality, 19, 37, 45, 52, 55, 91–96, 98–109
Religion/religious, 4, 9, 58–61, 85, 86
Revolution, 157–159, 178, 179
Rhees, R., 1, 2, 22, 23, 92, 93, 105, 106, 132, 142–144, 164–167, 178
Richter, D., 10, 62, 72, 75, 76, 78, 83, 84, 88
Ridge, M., 113, 177
Ripstein, A., 97–99, 103
Rorty, R., 41, 44
Rule, 4, 13, 51–53, 55–57, 60, 61, 64, 65, 89, 94, 101, 116, 164, 176, 177, 180
Rummens, S., 67, 89, 109, 129, 159, 178

S
Schroeder, S., 14, 53, 54, 57, 76, 114, 121, 123, 129
Science/scientific, 24, 40, 41, 43, 103, 108, 157, 158
Scientism, 3, 179
Sense, 11, 33, 51, 73, 91, 114, 132, 180
Seriousness, 138, 166
Shoemaker, S., 125
Singer, P., 153, 155–160, 171
Sinnott-Armstrong, W., 64–66
Sorensen, R., 162
Speciesism, 158
Starkey, C., 116, 121, 128
Strawson, P.F., 23, 154, 160
Stroud, B., 147

T
Taylor, C., 170, 178
Teichmann, R., 97
Testimony, 179
Theory, 3, 5, 6, 22, 73, 89, 91, 93, 101, 105–107, 115–118, 134, 149, 162, 167–169, 178
Therapist, 15, 31–33, 36, 40, 45
Therapy/therapies/therapeutic, 4, 6, 14–16, 31–46, 49, 56, 57, 176
Tilghman, B.R., 1, 87
Truth, 6, 19, 23, 65, 86, 91–96, 98–107, 109, 153, 159, 170, 176

U
Use, 1, 2, 33, 49, 50, 71, 94, 114, 128, 145, 154, 177

V
Value, 10, 21, 22, 60, 61, 87, 89, 102, 113, 140, 155, 181
View from nowhere, 93, 96, 97, 102, 103, 105, 107–109
von Wright, G.H., 64

W
Walzer, M., 153, 170
Watkins, M., 121, 122
Wiggins, D., 97, 99–102, 114, 116, 117, 126, 128, 178
Winch, P., 1, 178, 180
Wisnewski, J., 21, 50, 114, 116, 120, 128
Wright, C., 21, 100, 114, 118, 121, 124–128

FSC
www.fsc.org
MIX
Papier | Fördert gute Waldnutzung
FSC® C083411

Kolibri 360 GmbH
Ferdinand-Jühlke-Straße 7
99095 Erfurt, Deutschland
produktsicherheit@kolibri360.de